EMPLACEMENT MECHANISMS OF NAPPES AND THRUST SHEETS

Petrology and Structural Geology

VOLUME 9

Series Editor:

ADOLPHE NICOLAS

*Department of Earth and Space Sciences,
University of Montpellier, France*

Emplacement Mechanisms of Nappes and Thrust Sheets

by

OLIVIER MERLE

*Department of Earth Sciences,
University of Clermont-Ferrand,
France*

KLUWER ACADEMIC PUBLISHERS
DORDRECHT / BOSTON / LONDON

A C.I.P. Catalogue record for this book is available from the Library of Congress.

ISBN 0-7923-4879-6

Published by Kluwer Academic Publishers,
P.O. Box 17, 3300 AA Dordrecht, The Netherlands.

Sold and distributed in North, Central and South America
by Kluwer Academic Publishers,
101 Philip Drive, Norwell, MA 02061, U.S.A.

In all other countries, sold and distributed
by Kluwer Academic Publishers,
P.O. Box 322, 3300 AH Dordrecht, The Netherlands.

This is a translation of the original work
in French
"Nappes et Chevauchements" published by Masson.
Translated by G. Keaney.

Printed on acid-free paper

All Rights Reserved
© 1998 Kluwer Academic Publishers
No part of the material protected by this copyright notice may be reproduced or
utilized in any form or by any means, electronic or mechanical,
including photocopying, recording or by any information storage and
retrieval system, without written permission from the copyright owners.

Printed in the Netherlands

CONTENTS

Preface   vii

1   The Concept Of Thrusting   1
1.1 Nappes, Overthrusts And Fold-Nappes   1
1.2 Thick And Thin-Skinned Thrusting   8
1.3 Ramp And Flat Geometry   12
1.4 The Concept Of Displacement   18
1.4.1 Continuous And Discontinuous Deformation   18
1.4.2 Displacement Gradient   21
1.4.3 The Case Of Fold-Nappes   22
1.5 The Measurement Of Displacement   23
1.5.1 Rule And Hypothesis   24
1.5.2 Calculation Of Shortening   25
1.5.3 Balancing   27

2   Background   29
2.1 Historical Overview Of The Discovery Of Nappes   29
2.2 The Question Of Emplacement Mechanisms   41

3   Mechanics   49
3.1 The Mechanical Paradox   49
3.2 Resolution Of The Mechanical Paradox   54
3.2.1 Hypothesis Of Fluid Pressure   54
3.2.2 Hypothesis Of A Low Viscosity Basal Layer   57
3.2.3 Hypothesis Of Gravitational Spreading   61
3.2.4 Hypothesis Of The Triangular Shape   65

3.3	Composite Models	72
3.4	The Role Of Erosion	74
3.5	Classification Of Emplacement Mechanisms	78

4 Kinematics 81

4.1	The Kinematics Approach	81
4.2	Theoretical And Experimental Strain Patterns	84
4.2.1	Strain Factorisation	84
4.2.2	Rear Compression	89
4.2.3	Ductile Gliding	91
4.2.4	Gliding-Spreading	92
4.2.5	Spreading	93
4.2.6	Extruding-Spreading	99
4.3	Displacement Trajectory	102
4.4	Folding	106
4.4.1	Active Folding	106
4.4.2	Passive Folding	109
4.5	Emergent Ramp	111
4.6	Lateral Ramps	117
4.6.1	Bow And Arrow Rule	117
4.6.2	Wrenching Component	118

5 A Few Examples From The Alpine Chain 123

5.1	The Internal Zone: Crustal Stacking And Basement Nappes	123
5.2	The Helvetic Nappes: Extrusion, Spreading And Ductile Gliding	126
5.3	The Pre-Alps: Rigid Gliding	126
5.4	Molasses: Testimony To Crustal Thickening	128
5.5	The Jura: A Fold And Thrust Belt	130
5.6	The Southern Alps: Rear Compression	133

References	135
Index	151
Legend for Geological Cross Sections	159

Preface

Nappes and overthrusts are the most representative geological structures in mountain chains. The issue of their emplacement mechanisms and of the driving force of these displacements is a major problem in tectonics which interests, for near to a century now and not without harsh controversies, a significant proportion of structural geologists and geoscientists who work in the field of rock mechanics. This book attempts to give a clear and didactic synthesis of the current knowledge of the concept of thrusting, principally by tackling two approaches, mechanics and kinematics, which have proposed some solutions to this problem.

At first (Chapter 1), the notions of thrusting are defined, with the most recent terminology and the most important geometric aspects. This introduction to the geometry of thrusts is logically followed by the presentation of their problem; the issue of the emplacement mechanisms (Chapter 2). Let us note in passing that the formulation of the concept and the presentation of its problem are associated historically, which justifies presenting them in the historical framework of this discovery before tackling the different solutions and mechanical hypotheses. These are detailed in Chapter 3 by following a chronological progression, and emphasising the divergences and oppositions between different models so as to cover them fully. The chapter on the kinematics (Chapter 4) then returns to the type of data which can be collected in the field, by clarifying the relationships between displacement and internal strain. To establish a link between the mechanical models and the kinematical models is one of the primary objectives of this chapter which demonstrates in particular that data on the kinematics collected in the field can be used to test the validity of mechanical models. In effect, each mechanical model is associated with a specific internal strain which can be defined and quantified: to find a strain pattern in the field in an allochthonous unit in return validates the mechanical model proposed for the emplacement of this unit. Finally, a synthesis of the emplacement mechanisms described during the preceding chapters, by briefly presenting nappes and overthrusts on a cross-section of the Alpine mountain chain, allows these mechanisms to be located in their natural environment (Chapter 5).

This book on the emplacement mechanisms of nappes and thrusts aims above all to be a synthesis of the knowledge accumulated principally in the field of mechanics and kinematics. To avoid the impression of a 'complete science' which is the failing of this type of summary, the various sources which constitute the base of my documentation are for the main part given as the text progresses, with the divergences and oppositions between authors abundantly emphasised. Thus through the list of references, this work will allow the interested reader to go further, thanks to the numerous bibliographic paths which are open.

Olivier Merle
March 1997

CHAPTER 1

1 THE CONCEPT OF THRUSTING

1.1 Nappes, Overthrusts and Fold-Nappes

The concept of thrusting was defined towards the end of the nineteenth century. To get to the very heart of the matter, in this chapter essentially devoted to the most elementary definitions, it will be useful to turn to one of the most active supporters of what at the time was called "the theory of great nappes". Pierre Termier defined thrusts in the following way during a lecture given at Liège in 1906:

"A nappe, is a package of earth which is not in its place, which rests on a chance substratum, on a substratum which is not its original substratum" (Termier 1922, p.53).

This definition which is still perfectly correct, retains the key notion of *displacement*, of one unit or of several geological units on a thrust surface. This displacement is responsible for the formation of one unit named an allochthon which rests on a substratum called an autochthon bounded by a surface of structural discontinuity, that is to say a thrust fault which interrupts the normal stratigraphic or structural succession of the autochthonous terrain. In mountain chains, these displacements are the rule and an orogenic zone is, almost by definition, a region of nappes and thrusts. From one thrusted unit to another the observed displacements are of very variable size but can reach several tens of kilometres and even exceed a hundred kilometres.

In some of the geological literature, overthrusts are defined as the faulted superposition of one unit of a certain age resting on geological units of a more recent age. However, it is also conceivable that if the overthrust is of a large displacement, a nappe composed of units of a certain age can finally end its course on deeply eroded more ancient rock. This is Pierre Termier's "chance substratum". To define thrusts in this way is to confuse what is only a criteria of definition (amongst others) with the actual concept itself.

Three names, of a purely geometric nature, describe geological structures related to the concept of thrusting; nappes or thrust sheets, overthrusts and fold-nappes.

The term *nappe* or *thrust sheet* is used to describe huge displacements (several tens of kilometres) on a surface close to the horizontal (Fig. 1). In this case, the rock unit is displaced the right way up, without the formation of an inverted limb at the scale of the nappe, even if some smaller scale folding is frequently observed. The word "nappe" is of French origin meaning tablecloth and reflects the three-dimensional form of these geological structures which overlay a large surface whilst having a relatively moderate thickness. It is equally implicit that the nappes are independent of their homeland of origin, that they are in some way structurally detached from their root, and that these roots are so heavily tectonised that they are identifiable only with difficulty.

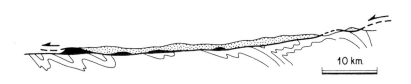

Figure 1 - Example of a nappe; the Parpaillon nappe in the western Alps (modified after Kerckhove 1969). In black; tectonic thrust outlier (cf p.14).

On the other hand, in general the term overthrust is used for more moderate displacements, from tens of meters to several kilometres (Fig. 2). In this case, the thrust fault is rooted at depth in a reverse fault whose angle of dip is generally between thirty and sixty degrees to the horizontal. The classic image of an overthrust is one of two fault blocks, initially adjacent, and where the overlapping of one on top of another has not totally altered or only partially obscured the original lateral structural arrangement. It is obvious from this that nappe and overthrust are two names which refer to the same phenomenon but represent two different stages of evolution. A nappe, in its initial stage necessarily passes by the stage of an overthrust before becoming an independent unit. It is therefore not surprising that the same geologic structure can be called at the same time nappe and overthrust, where the scale of the displacement is significant without being enormous, or where there is combined thrust fault rooting at the rear and multi-kilometric displacement along a horizontal surface towards the *front* (e.g. in the western Alps, one speaks indifferently of the Digne overthrust or the Digne nappe). At another scale, the faulted superposition of one slice of continental or oceanic crust onto another can be referred to equally under the term of *crustal overthrust*. In this case, the displacement along the thrust surface can be considerable (several hundred kilometres) but it is the image of two adjacent crustal slices thrusting one upon another which forces the use of the term overthrust in spite of significant displacement.

On the other hand, the term *fold-nappe* refers to structures whose geometry is sensitively different (Fig. 3). This name refers to huge multi-kilometric folds layered almost to the horizontal and whose highly deformed reverse limbs are of a considerably reduced thickness. Historically, at the end of the nineteenth and the beginning of the twentieth century this term of fold-nappe renowned much importance in the central and western Alps, amongst the French-Swiss group of researchers (comprised of Bertrand-Schardt-Lugeon-Termier). This is because they estimated that a nappe was derived from an overfold whose amplitude caused a tectonic rupture either along the axial plane or along the highly deformed inverted limb, a rupture which propagated and became a thrust surface (Fig. 4). It can be seen, in effect, that an overfold can exacerbate itself by thrusting over large distances of many kilometres, or that a fold-nappe can transform itself into a real nappe following the disappearance, through rupture, of the inverted limb. Nevertheless, even if these developmental relationships are possible, they are not the rule, and it is useful to recall that the school of researchers investigating the geology of Scotland (Peach-Horne-Clough-Geikie), during the same period were in opposition about this point in relation to the geologists from mainland Europe, by defending the hypothesis that the cause of all nappes and thrusts was due to a simple reverse fault without folding.

It should be noted immediately, and this will be returned to at length in the chapter devoted to the mechanics of thrust tectonics, that this terminology of thrusting relies on a purely geometric description of the structures and makes no hypothesis as to the emplacement mechanism. It remains useful to characterise a thrust defined by a geological map and a cross-section. On the other hand, this terminology becomes inadequate as soon as the nature of the forces causing the displacement of the thrusted units are considered.

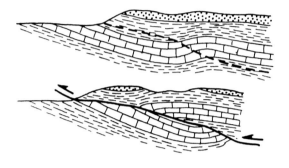

Figure 2 - Formation of an overthrust (modified after Hayes, 1891).

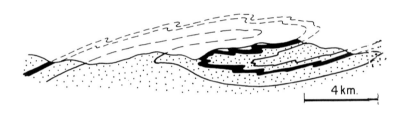

Figure 3 - Example of a fold-nappe in Galicia (Spain) (modified after Matte, 1968).

It is equally important to indicate that this classification is applied above all to thin-skinned thrusts situated in relatively shallow crustal levels, that is to say in the external zones of orogenic belts. Generally it occurs in a different way in internal orogenic zones where nappes form and deform at depths greater than fifteen kilometres. Again, we can turn towards Pierre Termier who, at the same lecture in Liège in 1906, defined a formula of the tectonic style of nappes in the most highly metamorphosed zones of orogenic chains:

"In a thrust region, I want to say in a region formed by the stacking of nappes, a lenticular shape is the rule" (Termier 1922, p.58).

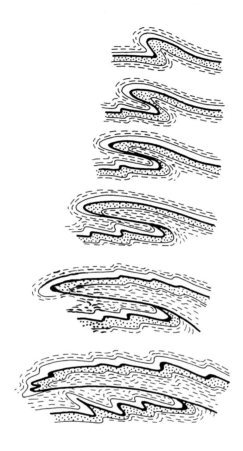

Figure 4 - Section showing the evolution from an overfold to an overthrust by rupture along the highly deformed inverted limb (modified after Heim, 1919-1922).

In effect, it is the absence of the continuity of thrusted units which is the most striking in the analysis of geological maps of these intensively deformed regions, the absence of continuity represented on geological cross-sections in the shape of massive *lenses* (Fig. 5). Elevated temperature, which increases the ductility of rocks, combined with the remarkable intensity of the deformation, causes large scale boudinages which isolate the most mechanically resistant material within the most ductile units. These lenses are often of very varied nature (orthogneiss, sediments of continental and oceanic origin, slices of oceanic crust and even rocks of mantle origin) and reveal the appalling complexity of the displacements achieved in the internal zones of orogenies. This leads us to be more specific about an important point: whilst the emplacement mechanisms of nappes are mentioned, both in popularising work and in research articles, it is almost always nappes and thrusts emplaced in shallow crustal conditions which is considered. In the internal zones, nappes are deformed and metamorphosed at too great a depth in order that the issue of displacements has no importance other than the one of the general kinematics of the tectonic event responsible for these huge deformations. To study the emplacement mechanisms of nappes within the internal zones returns to the study of the actual orogen itself and not any particular unit as it is possible to do it in less metamorphosed zones. For this reason, this book, whilst it treats the question of emplacement mechanisms, refers implicitly to displacements which occur at (or not far from) the surface, and in any event at depths which do not exceed approximately ten kilometres. Likewise, crustal thrusts, such as the obduction of oceanic crust onto a continent (example of the Oman ophiolite; Michard et al. 1989, Nicolas 1989) do not enter into the framework of this book devoted to the emplacement mechanisms of shallow crustal allochthonous units.

Figure 5 - Example of lens structures in the internal zones of the French Alps (modified after Philippot 1989).

We have already had the opportunity to indicate a classic criteria to recognise thrust structures, of one geological unit of a certain age overriding a terrain of a more recent age. This is the most obvious geometric result of a thrust as previously defined (Fig. 2). A similar type of criteria is applied in the internal zones where highly metamorphosed units overlay less metamorphosed units following an overthrusting even later than the regional metamorphism.

When a displaced unit rests on a more ancient bedrock it becomes more difficult to demonstrate the existence of the thrust. A time break in a stratigraphic sequence is most often detectable at the level of the thrust fault but one such sedimentary hiatus is not a conclusive argument because it can be associated with the sedimentary history within the field of paleogeography. The recognition of a thrust becomes more convincing where the overlying rocks have, in relation to the neighbouring rocks, some unrelated or exotic lithological and sedimentary characteristics which indicate a distant paleogeographic origin. Equally the presence of a reverse metamorphism under the fault contact of the nappe can be indicated, a metamorphism which is marked by a reversal of its gradient where the fault contact is crossed, although this is very difficult to demonstrate in practice. Such reversals of the metamorphic grade are understood to be the result of a heat released at the level of the thrust fault by the deformation of the basal layer (Aprahamian and Pairis 1981).

In the last instance, it is the analysis of the internal strain within the thrust fault which most often allows the ambiguity to be lifted, when this ambiguity remains. The base of nappes and thrusts are, in the majority of cases, the site of an intense deformation which contrasts sharply not only with the deformation recorded in the principal body of the thrusted unit but equally with the deformation observable in the autochthon. This deformation is naturally not always identical and depends on the pressure-temperature conditions present at the base of the thrusted unit at the time of the principal displacement. In a very shallow environment, the deformation is brittle and the thrust fault is most often associated with cataclastic rocks or with gouges (e.g. Brock and Engelder 1977). Striations visible on certain movement planes, allow the directions and sometimes the general sense of displacement to be determined. On the other hand, when the temperature is sufficiently elevated, deformation becomes ductile and a mylonitisation of rocks is evident all along the length of the thrust surface (e.g. Rhodes and Gayer 1977). Analysis of the elements of the internal strain (schistosity, stretching lineation, shear criteria, sheath folds...) again allows the sense and the direction of the displacement to be clarified better but also, in the case of the thrust planes, the functioning of the observed contact through reverse faulting.

Through these very general considerations on the deformation within the fault contact, we touch on the importance of the internal strain produced in the thrusted unit during the displacement. This internal strain possesses the characteristics which, in themselves, can be considered as some criteria to recognise a thrust structure. Moreover, this internal deformation can equally be significant for a particular emplacement mechanism. We will return in detail to these questions in the chapter devoted to the kinematics.

1.2 Thick And Thin-Skinned Thrusting

Retaking our earlier example of the overthrusting of one slice of continental crust onto another, it is geometrically clear that the major consequence of such a thrust is, in the vertical, a considerable crustal thickening (Fig. 6). The reverse fault causing a crustal overthrust can affect almost the entire thickness of the crust and can double the crustal thickness at the level of the overthrust. We will not approach here what corresponds to the thermo-mechanical evolution of the crust following such crustal thickening. Let's say only that the increase in temperature due to the thermal relaxation of the segment of thickened chain, in modifying the rheological properties of the deepest zones controls the kinematics of subsequent deformations. It is therefore all the history of the orogeny which is dependent on the thrusts and the crustal thickenings.

In presenting things in a slightly different angle, it can be said that the thickening becomes non negligible at the crustal scale, under the sedimentary cover, where the basement is equally involved in the reverse fault and comes to overthrust the overlying sedimentary beds. The process of crustal thickening is therefore indistinguishable from what is referred to under the name of thick-skinned thrusting, a name which indicates simply that the underlying basement is involved in the overthrusts, and that therefore as a consequence the deformation and the displacements occur reasonably at the scale of the entire crust.

However, in sedimentary cover sequences, it is often observed that a reverse fault, where the dip varies between thirty and sixty degrees, is rooted horizontally at depth in a particular level which it does not cross-cut, thus leaving the underlying levels unaffected by any deformation (Fig. 7). This type of curved reverse fault, of concave shape, is called *listric*. In this case, the basement is not involved and the name *thin-skinned thrusting* is used to indicate that at the crustal level the thickening is negligible. It is easily understood that a thin-skinned thrust be localised in the most external parts of mountain belts, which means far enough from the source of the stresses causing crustal deformations which occur in the more internal zones of orogenies. In other words, thin-skinned thrusts occur in the zones where the increase in temperature is negligible and where lithology remains the determining factor of the mechanical properties of the rocks. It is for this reason that this type of listric reverse fault does not root itself in any lithological horizon, but almost always within a lithological level of weak mechanical strength such as black schists, marnes or evaporites (e.g. Davis and Engelder 1985, Marcoux et al. 1987). This level where displacement is concentrated, tangentially to the stratigraphic marker layer, relates to what is called a *décollement horizon* which bounds the overthrust unit at its base. Again, in the external zones of mountain chains, the basement-sedimentary cover boundary is a major mechanical interface on which thin-skinned thrusts come to root themselves naturally, and all the more so if the sedimentary sequence of rocks overriding the basement is a horizon of weak mechanical resistance as is the case, for example of the Triassic gypsum of the Alpine chain. By way of example, the Jura is particularly representative of the displacement of a sedimentary cover sequence over an undeformed basement along a salt-bearing type of decollement horizon.

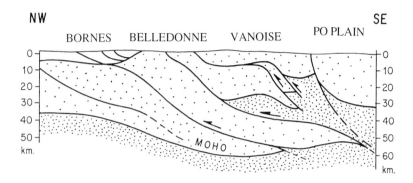

Figure 6 - Crustal scale thrust faults, example from the French-Italian Alps (modified after Tardy et al. 1990).

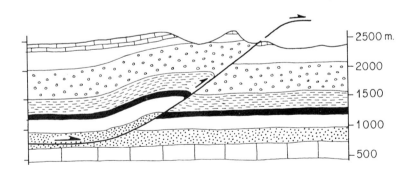

Figure 7 - Rooting of a thrust fault into a decollement horizon: example of the Ruby Inn Thrust in Utah (U.S.A.) (modified after Lundin, 1989).

The reverse fault along which the thrusting occurs strictly said, and which emerges to the surface in causing a local low amplitude thickening is referred to by the name of *emergent ramp* (or *frontal ramp*). In a more general way, a thin-skinned thrust can express itself by a series of *ramps* (high angle reverse faults within rigid or competent levels) and *flats* (horizontal displacement along ductile or mechanically incompetent layers) within the sedimentary succession. Contrary to widespread opinion amongst European geologists, the observation and interpretation of ramps and flats according to the deformability of the sedimentary layers are very old. Already in 1891, Willard Hayes was giving remarkable illustrations of this in the Appalachian mountain chain in the United States (Fig. 8) (cf also Rich 1934). We will return to this geometry of ramps and flats when we approach the terminology of thin-skinned thrust tectonics later in this chapter.

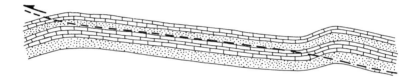

Figure 8 - Potential localisation of the ramps in competent layers and of the flats within incompetent layers (modified after Hayes 1891).

It is useful to shed light from here onwards on a particular aspect of the mechanical analysis of nappes and overthrusts with which we will be confronted in the presentation of mechanical models in the third chapter. Until a relatively recent period, mechanical models of nappes and overthrusts would always concerns the emplacement mechanism of a particular unit, studied separately, isolated and presented as an ideal case with its whole series of hypotheses on the mechanics relating to the nature of the forces and the rheological properties of the thrusted material. More recently, studies have concerned the mechanical analysis and the interpretation of all of the structures related to the overthrust observed in a foreland region of a mountain chain. In this way, the Jura can be analysed mechanically in its entirety without particular reference to the detail of the structures associated with the displacement of the sedimentary cover sequence on its rigid basement (cf for example the mechanical analysis of the Jura by Mugnier and Vialon in 1986). Therefore only the general geometric characteristics of these *fold and thrust belts* are taken into account (Fig. 9) to define the mechanical parameters which control the deformation and displacement of these thin-skinned thrusts. This change of scale is the work of the Americans (e.g. Elliott 1976a, Chapple 1978, Davis et al. 1983) who, with the Appalachian and the Rockies fold and thrust belts, have some exceptional field study areas perfectly characterised on the geological map. It is important to retain this point of scale in mind to avoid confusion during the presentation of models of the mechanisms of the emplacement of nappes and overthrusts.

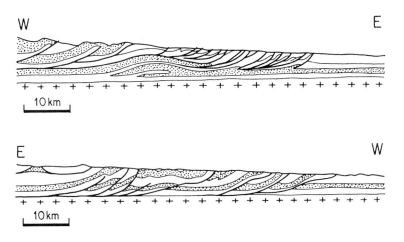

Figure 9 - Triangular shape or wedge-shape of the fold and thrust belts in the external zones of mountain chains. Above; Canadian Rockies (modified after Bally et al. 1966), below; northern Appalachians (after Roeder et al. 1978).

Let's look immediately at the idealised geometry, shared amongst almost all fold and thrust belts;

1 Fold and thrust belts most usually have a general *wedge shape* related to opposite dips of the upper surface slope and basal slopes. The dip of the upper surface slope (several degrees) is inclined towards the foreland whilst in reverse the dip of the undeformed bedrock (equally several degrees) is oriented towards the internal zone of the orogenic mountain chain. From this fact, the belt is thicker at the rear in the hinterland than at the front.

2 Fold and thrust belts almost always comprise at their base a particularly deformable decollement horizon along which the principal displacement is localised.

3   Within the belt, structures relate to the imbricated thrusts whose principal geometry is that of listric reverse faults with ramp and flat development associated with lithological variations within the sedimentary sequence.

4   Displacement dies out towards the front of the belt.

5   Deformation is more significant towards the rear of the belt.

If, from here onwards, we concentrate on fold and thrust belts, it is because their study has recently contributed to reinforce strongly the field of investigation of the mechanics of nappes and overthrusts. We will come back to this.

1.3   Ramp and Flat Geometry

The study of fold and thrust belts in the USA (the Rocky Mountains and the Appalachians) has led to a comprehensive terminology of the geometry of overthrust sequences (e.g. Boyer and Elliott 1982, Butler 1982, Jones 1987, Graham et al. 1987). The terminology is principally based on the idea of a *ramp-flat* geometry which has already been mentioned (cf p. 10). The trajectory of thrust faults, viewed in cross-section, often relates to a sequence of sub-horizontal surfaces (or flats) in incompetent layers and steeply sloping surfaces (or ramps) which climb up across the most mechanically competent layers. The geometric relationship between ramp-flat and competent-incompetent layers has been known for a long time (Hayes 1891) and has been confirmed by small-scale modelling performed in laboratory experiments (Ballard et al. 1987). This geometry is probably due to the fact that high angle reverse faults initially form in competent layers. The dip of the fault plane (approximately 30 degrees) corresponds to the value which would be expected if the maximum compressive stress is sub-horizontal. For reasons of compatibility, these ramps cannot be effective as displacement planes until they are connected by the 'flats' in the incompetent layers. This occurs by the propagation of different thrust faults in the overlying and underlying flats until a continuous series of flats between ramps is achieved (Eisenstadt and De Paor 1987) (Fig. 10). It is the complexity of this particular geometry, typical within sedimentary sequences where mechanical contrasts between stratigraphic layers is sometimes markedly strong, which causes the abundant terminology associated with thin-skinned thrust tectonics.

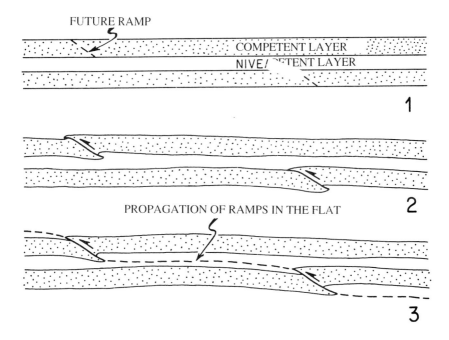

Figure 10 - Formation of ramps and flats. The ramps initially form in competent layers (1 and 2) and are then connected to each other by horizontal propagation of the thrust fault along the 'flats' in the incompetent layers (modified after Eisenstadt and De Paor 1987).

Reverse Fault. The angle of dip of a reverse fault is often characteristic of the ramp-flat geometry. Dipping at $60°$ to the surface where it emerges (emergent ramp) it is often rooted horizontally at depth in an incompetent layer (flat) (Figures 7 and 11). Ramp propagation towards the surface sometimes results in the development of a fold in the hangingwall in order to accommodate displacement on the ramp. In the case of a *fault-propagation fold* (Suppe 1985, p. 350) the ramp does not migrate up to the topographic surface but terminates along the axial plane of the syncline associated with the anticline which accommodates the fault displacement (Fig. 11). In the opposite situation, where the ramp is connected either to the surface or to the upper flat, the passing over from the ramp to the horizontal surface creates a fold in the hangingwall of the thrust. Apart from mechanical reasons, the hangingwall fold or *fault-bend fold*, (Rich 1934) is necessary for kinematical reasons. It is therefore a passive fold, which may form even in the absence of a mechanical contrast between the different layers involved (Fig. 11).

Often an inclined thrust plane may develop in the opposite sense branching off a ramp. In mechanics, the geometrical arrangement of antithetic and synthetic faults is known as *conjugate faulting*. The fault dipping in the opposite direction is called a *backthrust*. Where a backthrust truncates (or is truncated by) another conjugate fault situated immediately behind it, this forms a *triangle zone* (Elliott 1981).

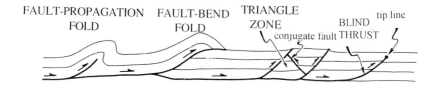

Figure 11 - Structures associated with the formation of reverse faults in thrust belts.

Imbricate Structure. This concept was initially introduced by Suess in 1883 by the term *Schuppenstruktur* and later translated by Hobbs (1894) as *imbricate structure*. This term describes a series of regularly spaced and equally-dipping thrust faults. All thrust faults within an imbricate structure originate from a common horizontal *sole thrust* or *floor thrust* at depth, from which reverse faults branch off to form individual thrust units (Fig. 12). Reverse faults which do not propagate to the topographic surface are termed *blind thrusts* (Thompson 1981) (Fig. 11). This feature is particularly common towards the foreland front of a thrust belt (Morley 1986) and indicates that the displacement of an overthrust is not infinite: the line which defines this limit of displacement is called the *tip line*.

When each new thrust fault develops in the footwall of an older thrust, this sequence of emplacement is called *in sequence* or *piggy-back thrusting*. On the contrary, where each new thrust develops in the hangingwall of an older thrust, the sequence is termed *out of sequence* (Fig. 12). Piggy-back thrust sequences are the most frequently occurring in nature, probably because the energy required to initiate this type of propagation is less than in the case of out of sequence thrusting (Mitra and Boyer 1986).

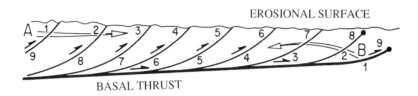

Figure 12 - Imbricate structure and sequential development of thrusts in a piggy-back sequence (foreland propagation; arrow A and numbers indicate the order of development of thrusts). Out of sequence thrust stack (propagation of thrusts in the hanging wall; arrow B and numbers indicating the order of development of thrusts).

Duplex. Where an imbricate structure is bounded above by a *roof thrust* and below by a floor thrust, this particular imbricate structure is termed a *duplex* (Dahlstrom 1970, p. 352) (Fig. 13). Such structures may exist at all scales, ranging from a kilometre to a centimetre (Tanner 1992). Between the roof and floor thrusts, the faults bounding each imbricate slice have a sigmoidal shape and are parallel to the roof and wall of the duplex. The smallest component of the duplex, bounded between two of these sigmoidal faults are fault blocks called *horses* (Fig. 13). The internal stratification of each horse is overall parallel to the two sigmoidal bounding faults, typically corresponding to a characteristic S or Z shaped geometry (Fig. 13). In the duplex roof and wall, the internal bedding or stratification can be relatively gentle and even a horizontal stratigraphic layer may form the roof (or the wall) of the duplex. In most cases, the dip of the horses conforms to the general vergence, i.e. to the overall sense of displacement of the duplex. A *hinterland dipping duplex* is most often associated with a piggy-back type thrust sequence (Boyer and Elliott 1982). On the other hand, an out of sequence duplex can result in the dip of horses being non-conformable to the general vergence of the *duplex*. In the case of a *foreland dipping duplex*, the internal stratification is not always parallel to the sigmoidal faults which bound the horses and may even become uplifted and inverted into an opposite orientation (Fig. 13). The continuation of tectonic shortening can cause the deformation of the duplex by the overthrusting of one horse onto another. These thrust faults rapidly jam and the whole structure folds forming an antiformal stack (Fig. 13).

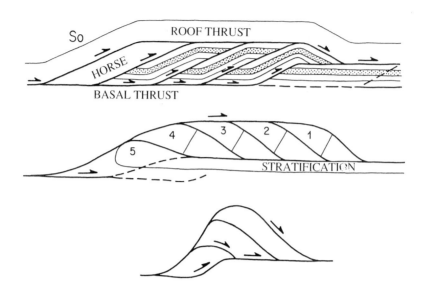

Figure 13 - The elements composing a duplex (modified after Boyer and Elliott, 1982). Above: hinterland dipping duplex. Middle: foreland dipping duplex. Below: antiformal stack formed by the uplift and folding of the duplex by successive activation of thrusts.

It goes without saying that this terminology, together with the numerous other descriptive terms associated with the concept of overthrusting familiar to field geologists, allows a comprehensive description of the majority of structures associated with nappes and thrusts. As the objective of this book is not to be a comprehensive description of overthrusting, a small glossary is proposed, however inevitably incomplete, of the most common established terminology which has not been discussed previously.

Diverticulation. Term by which Lugeon (1943) attempted to explain some faulted superpositions where in the vertical section, a stacking of formations of the same sedimentary series in a reverse sense of the normal stratigraphic sequence is observed. This superposition would result in sliding in quick succession of the different formations of the sedimentary pile. The upper units would become detached the first, starting with the highest unit, and would slide forwards until they become immobilised and are then caught up with by the lower sedimentary formations which have started to slide later. These formations end their course by overthrusting the units initially structurally higher, the lowest formation then being found at the top of the thrust structure while the highest formation is relegated to its base.

Tectonic slice. Term of imprecise meaning which generally describes a small scale slice of rock which is found in an allochthonous position within the largest assemblage. Used most frequently to describe a component of basement or of competent sedimentary material isolated in the largest scale thrust units.

Encapuchonnement. This term, introduced by Argand (1911), describes a particular geometrical arrangement where the front of a nappe is totally overlapped and enveloped by the autochthon or the underlying nappe. This arrangement occurs following folding of the underlying unit, which becomes folded over the front of the upper unit in the opposite sense of thrusting.

Window. Delimited zone, exposed by erosion within a nappe, of autochthonous rocks or a lower nappe.

Klippe. Allochthonous rocks, isolated and often of small scale, relating to the sole relic of a nappe which has disappeared through erosion. A klippe rests on a thrust fault close to the horizontal on the autochthonous unit or another nappe still preserved from erosion.

Thrust outlier. Portion of rocks detached from the authothon (or underlying allochthon) and dragged under the nappe during its passage on an erosional surface. It is supposed that these outliers represent ancient reliefs which hindered the displacement of the allochthonous unit.

Olistostrome. Final formation of a sedimentary basin showing chaotic blocks of all scales belonging to a nappe (or from several nappes) which rest on this formation. Each exotic block is called an olistolith. It is recognised that these blocks have become detached from the front of the nappe involved in the basin and have glided forwards to deposit in the final formation before the passage of the nappe permanently ends the sedimentation in the basin.

Parautochthon. Name for a rock unit situated under a sequence of nappes and where it is suspected, in spite of its stratigraphic and lithological continuity with the real underlying autochthon, that it has also been slightly displaced at the time of the emplacement of nappes.

Backthrusting. Describes displacements, generally late stage in the history of the chain, whose vergence is in the opposite sense to the general displacement of the allochthonous units. In a mountain chain, these are therefore displacements towards the internal zones and not towards the external zones as is most often the case.

Cover substitution. This is said when a thrusted sedimentary cover sequence comes to rest on the uncovered basement as if it was the autochthonous cover sequence.

Synformal anticline. Anticlinal hinge of the front of a fold-nappe bedded beyond the horizontal and which appears as a synform overturned in the opposite direction to the overall displacement.

1.4 The Concept of Displacement

1.4.1 CONTINUOUS AND DISCONTINUOUS DEFORMATION.

The term 'displacement' is often poorly defined and the source of confusion. Yet it is of paramount importance regarding nappes and thrusts since it is the base of the definition of thrusting (cf introduction p. 1). The issue is to clarify the relative roles of rigid body translation and of internal or ductile deformation in the overall amount of displacement achieved. With this objective it is useful to recall two basic concepts and to see to what degree they help in the understanding of the problem of the measurement of displacement produced by nappes and overthrusts:

Rigid body translation. A body undergoes a rigid body translation when all points within a reference frame displace in exactly the same direction and by the same amount. Naturally, this implies that the object suffers no internal distortion or any type of rotation. It is equally clear that in this case the distance between the reference points does not change during displacement.

Internal or Ductile Deformation. Contrary to rigid translation, different reference points may change their relative positions during the course of a transformation. This shape change shows that the material has undergone distortion, or internal strain.

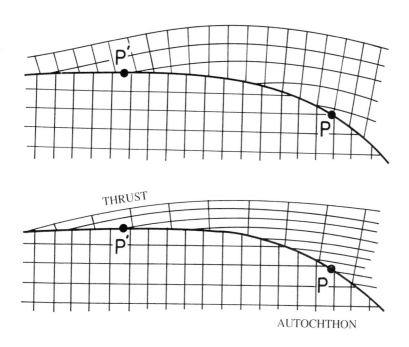

Figure 14 - Example of identical displacement (points P and P' are initially adjacent) produced by rigid translation (above) and by combination of rigid translation and internal strain (below) (modified after Ramsay 1969).

Without entering into the mathematical formulation of these two concepts, it should be noted immediately that displacement in geology can be the result of a rigid translation, of an internal strain or of a combination of these two components. With reference to thrust faulting, the displacement is localised along the length of the thrust plane regarded to be the major discontinuity. Following a thrust displacement, the distance which separates two points which were initially adjacent but situated on either side of the fault surface provide the best measurement of displacement achieved on a thrust fault. However, this does not prejudice the way in which displacement is accomplished and the same amount of displacement can be produced as well by a rigid body translation (Fig. 14 above) and by the combination of rigid body translation and internal or ductile deformation (Fig. 14 below). As an example of a combination of rigid translation and ductile deformation in the case of a hinterland compression, the internal strain, very intense at the rear, progressively diminishes towards the front which may be practically undeformed. As in the example of the Digne Nappe in the western Alps (Siddans 1979), displacement is achieved by a combination of rigid translation and internal strain at the rear of the thrust sheet, to a displacement achieved uniquely by rigid body translation at the nappe front.

Up to now, we have made the implicit hypothesis that the overthrust fault corresponds to a surface discontinuity, that is to say that displacement occurs along a planar fault surface of no given thickness and that the autochthonous and allochthonous units were in contact as two superposed blocks. However, displacement can occur by intense deformation in a basal low viscosity bedding layer (p. 57). In this case, displacement is accommodated by ductile deformation which may occur throughout the thickness of the low viscosity layer without any localised failure occurring between the autochthonous and allochthonous units (Fig. 15). The displacement is entirely

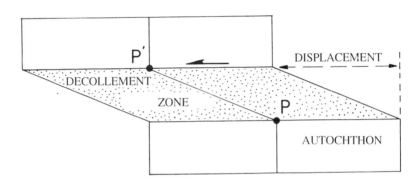

Figure 15 - Schematic representation of continuous deformation: displacement occurs by internal deformation in the decollement zone (the line PP' is initially vertical) without there being any rupture between the autochthonous and allochthonous compartments.

the result of internal strain within the level of the decollement zone (without a rigid translation component). This does not fundamentally change the quantification of the displacement. Therefore it is considered that it is the competent unit situated above the decollement zone which undergoes a rigid translation in relation to the autochthonous unit. The decollement zone, whatever its thickness, is likened to the fault which occurs at the base of the thrust during a discontinuous deformation. On the other hand, continuous and discontinuous deformation are not necessarily exclusive and displacement can occur firstly by continuous deformation followed by a discontinuous deformation afterwards, which means a rupture within the low viscosity basal layer. Likewise, a simultaneous combination of the two processes is possible, the continuous deformation of a mechanically weak layer not preventing discontinuous displacement at the base of this same layer. By way of example, one can indicate that of the seventy kilometres achieved by the Laksefjord nappe in Finland, only seven are assured by internal strain, the remaining sixty-three are achieved by rigid translation (Williams et al. 1984).

1.4.2 DISPLACEMENT GRADIENT.

The exact quantification of displacement is sometimes dependent on the internal strain in the allochthonous unit. Whether the deformation is homogeneous or not, the displacement is not uniform from the moment that the component of internal strain is no longer negligible. Therefore the amount of displacement varies along the thrust fault plane (Fig. 16). This displacement gradient is a function of the emplacement mechanism of the nappe or of the thrust sheet. In the case of a rear compression, the displacement is maximum at the rear and reduces towards the front (cf Fig. 39). Likewise, some gravitational gliding dies out towards the front and the sedimentary cover which has undergone sliding and displacement at the rear links up in continuity with the autochthonous series (for example, this is the case of displacements observed in the region of the Nice arc (Geze 1962)). On the other hand, other gravity emplacement mechanisms, such as gravitational spreading, create a maximum displacement in the front which reduces towards the rear. Extreme cases relate to displacement which dies out towards the front during rear compression or from gravitational gliding and which dies out towards the rear during gravitational spreading. Thus passing progressively from an allochthonous zone to an autochthonous zone, from the front towards the rear or from the rear towards the front according to the emplacement mechanism, progressive reduction of the internal strain is observed (cf Figs. 39 and 46). To simplify, it can be said that allochthonous units, depending on whether they are pushed from the rear or 'pulled from the front', have displacement gradients whose sense is opposite.

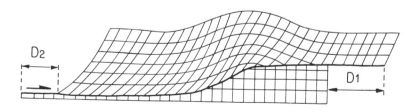

Figure 16 - The amount of displacement along the thrust fault plane (D_1 in front of the ramp and D_2 behind) can vary according to the internal strain in the allochthonous unit (modified after Elliott 1976).

The same reasoning is applied to the shortening which can vary from the rear to the front of nappes and thrust faults. It is stated that shortening increases towards the front of units emplaced by gravitational gliding (in reverse to the displacement) and towards the rear for units emplaced by rear compression. The calculation of the gradient of shortening from geological cross-sections can thus be used to determine where to propose an emplacement mechanism of the studied unit (e.g. Merle 1982). If the shortening occurs during the translation and is not inherited from a pre- of post-thrusting deformation, the information given by the shortening gradient about the emplacement mechanism can be used to demonstrate this, whatever the amount of displacement absorbed along the thrust fault surface even if the amount of displacement is possibly not quantifiable (which is the case for example for nappes which have been totally detached from their roots and displaced over several tens of kilometres).

Two borderline cases can be deduced:
1. Constant displacement along the thrust fault when displacement is produced entirely by rigid translation.
2. Displacement gradient along the thrust fault when the internal strain absorbs all of the displacement.

1.4.3 THE CASE OF FOLD-NAPPES.

Recumbent folds of several kilometres scale are formed by a mechanism which relates neither to a discontinuous deformation nor to continuous deformation along the base. Experiments on scaled models produced in the laboratory have demonstrated that fold-nappes form by frontal rolling which uncoils the reverse limb from the normal limb (Brun and Merle 1985, Merle 1986) (Fig. 17). During this process, the reverse limb progressively increases as the most forwards parts of the normal limb, by coiling in the frontal hinge which migrates continually forwards, finally settling in a reverse position. As soon as a point of the normal limb passes in a reverse position, it comes into contact with the autochthon and its displacement stops. The most simple image taking account of this process is the one of the rolling of the caterpillar tracks in front of a tank (C. J. Talbot, personal communication 1987). The maximum displacement produced like this way occurs right above the frontal hinge. By following the basal contact of the frontal hinge towards the rear, the amount of displacement regularly reduces and cancels out within a syncline which links up the autochthon with the allochthon. It is important to point out that the internal strain at the base of the allochthonous unit is relatively intense and not too different to what is observed when thrusting occurs by continuous deformation of a ductile basal horizon. We will see in the chapter devoted to the kinematics that gravitational spreading always produces a frontal fold-nappe and that therefore it is advisable to study the internal strain of the fold-nappes in the light of these spreadings.

1.5 The Measurement of Displacement

In nature, it is rare that a pre-thrust marker object can be observed which is cross-cut by the basal contact of a thrust. This can occur for example when pre-thrusting magmatic intrusions or veins are cross-cut by the thrust plane. When such a case occurs, it is obvious that the measurement of displacement therefore relates to the distance separating the two sections of this marker object on both sides of the basal surface. However, even in this favourable case, it should be remembered that the measured displacement may not be the same value updip or downdip along the thrust fault of the marker level, because of the gradient of displacement associated with overthrusting of the allochthonous unit.

Figure 17 - Formation of a fold-nappe by frontal rolling: three successive stages obtained from experiments on scaled models. The deformation of the initial squares allows the internal strain to be visualised.

In the absence of this type of marker structure, which means in the most frequent case, the most powerful method of measuring the displacement and the shortening is the method of balanced cross-sections (Dahlstrom 1969). This method consists of reconstructing the initial pre-thrust state from geological cross-sections and from strain analysis (Goguel 1952, Hunt 1957). The comparison between the reconstructed initial state and the observed final state (in reality interpreted in the field) allows the value of displacement at all points along the thrust surface to be estimated with relative accuracy. Equally the use of this method shows inconsistencies in the geological cross-section constructed in the field whose restoration to the initial state is often not possible, demonstrating how some hypotheses made in the field at the time of its construction are not valid. Therefore the cross-section of the final state can be corrected in accordance with inconsistencies which appear during the restoration of the initial state. When a good consistency between the two cross-sections is achieved, the final state cross-section is said to be balanced. As Gille Menard (1988) remarked, a balanced cross-section represents only a geometric solution amongst other solutions and is not necessarily correct, while a non-balanced cross-section is likely to be wrong. Therefore the method of balanced cross-sections goes well beyond the simple measurement of displacement and shortening since it allows geological cross-sections to be better constructed by allowing better hypotheses to be proposed for structures hidden at depth and therefore inaccessible to direct observation.

1.5.1 RULE AND HYPOTHESIS.

The rule to respect whilst making a reconstruction of the initial state is to consider only cross-sections constructed parallel to the displacement direction. Usually, structures such as folds and faults form perpendicular to the displacement direction but it is useful to keep in mind that it can be otherwise. To construct cross-sections perpendicular to the structures is therefore a hypothesis which must be justified by structural arguments.

Either way this rule of construction of the cross-section of the final state, there should be conservation of area in the plane of section between the initial and final states (Goguel 1952, p. 144). This hypothesis indicates on the one hand that there is a conservation of volume during the deformation (which is generally the case for rocks which have undergone diagenesis) and on the other hand that there has been neither lengthening nor shortening in the plane perpendicular to the section. This second point implies that the internal strain, if it exists, is plane strain and occurs in the plane of section. This hypothesis is not always confirmed and does not necessarily prevent the construction of balanced cross-sections. The method is therefore different, and more complex, since it is firstly necessary to quantify the three-dimensional deformation on a map before undertaking the restoration of the initial state (Gratier et al. 1989, 1991, Guillier 1991).

1.5.2 CALCULATION OF SHORTENING.

Three methods can be used to calculate the shortening from a cross-section constructed in the field.

Unfolding. This method consists of measuring by curvimetry the length (l_0) of marker beds in a geological cross-section. The comparison with the length after folding (l_1) allows the shortening from the initial state and the final state to be calculated [R (%) = 100 x (l_1 - l_0)/ l_0]. This method implies that the thickness and the length of beds are obliged to remain constant during the deformation. It is advisable to check that beds have not undergone internal strain. In the opposite case, it is necessary firstly to quantify this internal strain and to dissociate it from rigid rotation of beds associated with folding (Cobbold 1984, Woodward et al. 1986). On the other hand, it is equally necessary to quantify the discontinuous displacements on secondary thrusts which occur within the actual interior of the studied allochthonous unit. The offset of the marker beds from either side of the reverse faults allows the shortening associated with the functioning of these internal thrusts to be quantified.

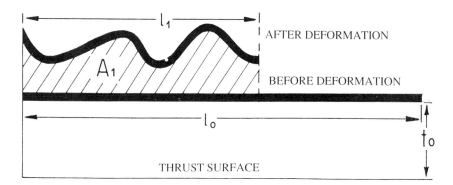

Figure 18 - The calculation of shortening from the equal area method (modified after Chamberlin 1910).

Excess section. This method consists of measuring the excess section A_1 for this bed after deformation, by evaluating the distance t_0 separating a given bed before deformation from the base of the thrust (Chamberlin 1910) (Fig. 18). The initial length l_0 is therefore given by the following equation:
$l_0 = (A_1 / t_0) + l_1$ (Gwinn 1970, Elliott 1977). The shortening is calculated as previously.

It should be noted that some sections of the marker bed can have been removed by erosion. In this case, it is necessary to reconstruct the geometry of this marker bed above the topographic surface on the geological cross-section in the zones where is has disappeared through erosion.

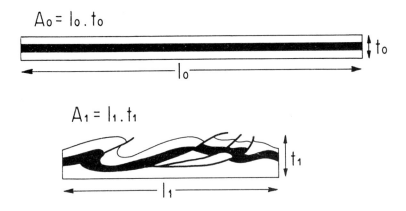

Figure 19 - The calculation of shortening from the total area method (modified after Hossack 1979).

Total area. The initial length l_0 can also be measured from the total area involved in the thrusting (Dennison and Woodward 1963, Kiefer and Dennison 1972, Hossack 1979). If the area of a section of length l_0 and thickness t_0 shortened into a section of thickness t_1 and length l_1 is identical before (area A_0) and after (area A_1) deformation, therefore it is known that $A_0 = A_1 = l_0 t_0 = l_1 t_1$. From it the initial length can be deduced: $l_0 = A_1 / t_0$ (Fig. 19). However, again, since the erosional surface indiscriminately cross-cuts folds and internal overthrusts, it is necessary to reconstruct the eroded parts according to the greatest thickness observed in the cross-section, so that the measured area is equal to the surface before deformation.

1.5.3 BALANCING.

The balancing of a geological cross-section is a relatively long process which is carried out by moving backwards and forwards between the cross-section of the final state (which is the starting material) and the initial state cross-section. The

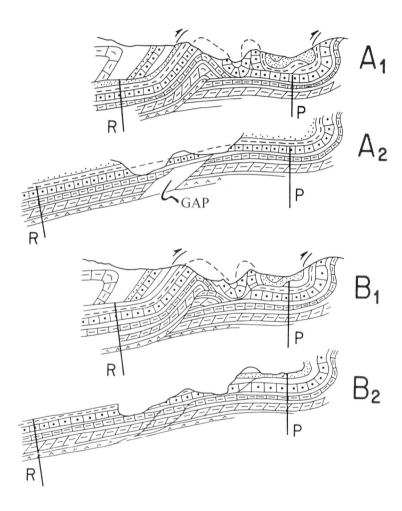

Figure 20 - Balancing of a geological cross-section. The geological cross-section initially constructed in the field (A_1) does not allow correct restoration to the pre-deformation state (A_2). Only a modification of cross-section A_1 allows a satisfactory restoration to the pre-deformation state (B_2). The corrected geological cross-section (B_1) is said to be balanced (modified after Menard 1988).

progressive elaboration of the cross-section of the initial state allows possible geometric inconsistencies of the final state cross-section to be demonstrated. This last point is corrected according to information which allows restoration to the initial state. As we have already indicated, the geometrical solutions which are used are not necessarily correct but aim to cancel out geometrical inconsistencies of the cross-section constructed in the field. In this case, the balanced cross-section proposes another interpretation of the hidden structure in the field, an interpretation which is more satisfactory than the initial starting cross-section.

It is advisable firstly to attach a vertical reference line, or a *pin-line*, in the autochthonous part of the starting cross-section. This line is fixed and allows the starting cross-section and the initial state cross-section to be superimposed onto each other. The technique consists of restoring the sedimentary sequence to the horizontal and a cancellation of the displacement along the faults by replacing initially adjacent stratigraphic marker beds into contact with each other. It is during this operation, which is made progressively from the most external zone towards the most internal zone, that the restoration sometimes reveals holes which violate the hypotheses of the method. The starting cross-section is therefore modified to match the central hypothesis of the conservation of surface in the plane of section (Fig. 20). When the restoration is finished, there must be complete conservation of the surface between the two states, then the cross-section of the final state is said to be balanced. Computer programs allow these operations associated with the balancing of cross-sections to be automated (e.g. Groshong and Usdansky 1986, Jones and Linnser 1986, Kligfield et al. 1986, Medwedeff and Suppe 1986, Endignoux and Mugnier 1990).

The interest in the technique, which is used above all in fold and thrust belts where internal strain is weak and the sedimentary sequences are easily identifiable, is not only to propose more viable geological cross-sections but also to quantify in a precise way the displacement at all points on the cross-section. Thus, on the balanced cross-section of Figure 20, the restoration allows the line R or all other lines positioned on the allochthonous part of the balanced cross-section to be followed along a course of retro-displacement.

CHAPTER 2

2 BACKGROUND

2.1 Historical Overview of the Discovery of Nappes

To attempt a historical perspective of the discovery of nappes and overthrusts requires that an important point is clarified straightaway; fault contacts resulting from the functioning of a reverse fault were well know during the early part of the nineteenth century and have never fuelled passionate controversies as was the case for nappes and major overthrusts. Reverse faults, in particular high angle reverse faults, are described as the result of a "lateral push" which brought about one block to overthrust its neighbour over a distance generally estimated to be several hundreds of meters, and at maximum, a kilometre.

Thus, from 1843, the Rogers brothers described asymmetric folds in the Appalachian mountains which evolve into reverse faults through the occurrence of rupture along the upright limbs of the anticlines. Not only is the importance of their discovery not contested but it is immediately acknowledged by the international community. Later, it is Logan in 1862 who suggests that the front of the Canadian Appalachians relates to the overthrusting which brought about the Lower Ordovician to structurally overlay the Upper Ordovician. The interpretation which he gives of this fault contact is simple and dispassionate, and leaves those within the geological world in no doubt about its validity.

"The Lower Palaeozoic strata, pushed up this slope, would then raise and fracture the formations above, and be ultimately made to overlap the portion of these resting on the edge on the higher terrace; after probably thrusting over to an inverted dip, the broken edge of the upper formation" (Logan 1862, p. 327).

In Europe, the whole of the structure of the French-Belgian Carboniferous coal-bearing basin is interpreted by incorporating the concept of major reverse faults by Gosselet in 1879. And, in Scotland, Lapworth in 1878 described some thrusts illustrated by using highly clear cross-sections (Figure 21). Most often, these reverse faults are supposed to relate to faulted folds (broken or disrupted folds) whose inverted limb would have disappeared through being faulted out during deformation:

"Each of these anticlinals is a faulted one, a portion of the inverted limb of each being lost" (Lapworth 1878, p. 265).

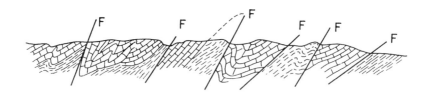

Figure 21 - Thrust faults by rupture along the limbs of anticlines (modified after Lapworth 1878).

If the existence of these reverse faults is not contested, it is essentially for two reasons; firstly because the thrust faults are generally steeply dipping, and then secondly because the envisaged or demonstrated displacements are relatively slight and only very rarely exceed a kilometre. But as soon as a researcher demonstrates an overthrust displacement of greater extent along a low angle thrust fault, suddenly opposition becomes very virulent. It was in this way that when Callaway (1883), still in Scotland, described some displacements which exceeded a mile (1.6 kilometres), scientific communication was carried out only in a very defensive way. Let's listen to the conclusion of his paper because it is significant of the type of discussion during that period and stages some of the big names of the scientific controversy that are going to be mentioned:

"If it be objected that the stupendous inversions and overthrows which I have described are improbable, I have only to reply that such effects are not uncommon in disturbed districts, and are familiar to geologists. I will only call attention to a remarkable example, copied from A. von Heim by Dr. A. Geikie, in his new Textbook of Geology, p. 518. On the left of the section 'schistose rocks, perhaps metamorphic Palaeozoic formations', with white Jura conformably underlying, rest upon the upturned edges of highly contorted Eocene strata" (Callaway 1883, p. 414).

To protect himself from the attacks of which he will be the target, Callaway hides himself not only behind two of the greatest names in geology of the period, but also behind a famous geological example which, moreover, divided and embarrassed the scientific community for a number of years. And yet, in the case of Callaway, he proposed displacements which remain of only moderate scale. But here situates itself the heart of the controversy which is going to turn geology upside down towards the end of the nineteenth century; to displace a rock mass over several kilometres along a fault contact close to the horizontal far surpasses what the geological community of the period could acknowledge. Such an hypothesis appeared to be all at once eccentric and mechanically impossible. But there is more. The prospect of giving up the theoretical foundations on the formation of mountain chains of that period, and the reinterpretations of the tectonics that this abandoning hints at, plunges geologists into a total (and unbearable) uncertainty. This apprehension must be understood; in effect what would remain of the tectonic outline of the Alps after the instigators of the new concept had reinterpreted this mountain chain? Practically nothing!

And yet, the example of which Callaway speaks, and what all the geologists know, is so demonstrably clear that it is almost inconceivable that fifty years have been necessary to explain it correctly. What is it?

In 1841, the Swiss geologist Escher von der Linth published the first results of his work on the Glaris region of the Alps[1]. In this paper, Escher demonstrated that on a surface of almost fifty kilometres in length and twenty five kilometres width, the Nummilitic (Eocene) occurs in the base of valleys and is overlain by Permian conglomerates and its overlying Liassic sedimentary sequence which defines the valley hilltops (Figure 22a). Not leaving the facts that he has shown, he prudently puts forward the hypothesis that this discontinuous succession is the result of an enormous upturning of the Permian onto the Tertiary

---

[1] Concerning the discovery of the nappes in the Alps, the reader is advised to read the fascinating work by Bailey (1935) just as the papers by Masson (1976, 1983) and Trümpy (1991).

sequence [2]. At the same time, he knew that he had no idea about the way in which such a thing could occur. Nevertheless, the quality of Escher's work is so remarkable, the arguments so precise and solid, that nobody contested the existence of the phenomenon. It is an enigma, and has convinced many of this faulted or abnormal superposition, as in the case of the Englishman Murchison, initially more than sceptical, but who returned to Britain totally convinced by the facts.

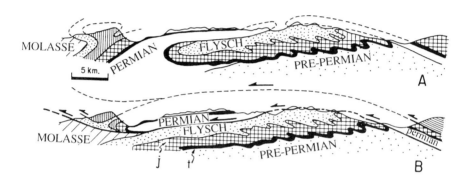

Figure 22 - Structure of the Glaris region of the Alps (modified after Bailey 1935). a) Heim's theory of the double fold. b) Bertrand's theory of one single thrust towards the north.

Escher did not like to write and his publications are rare. In the following years, he presented orally his interpretation of "kolossalen Ueberschiebung" to all those who came to visit him. This explanation which he obstinately refused to publish [3] nevertheless became famous throughout the whole world under the name of "the Glaris double fold". At the death of Escher in 1872, his student Albert Heim succeeded him to the Chair in Zurich. The latter rapidly became the

---

[2] For Escher, the thrust fault occurs "following a huge overlap or upturning of beds" (Folge einer kolossalen Ueberschiebung oder eines Umbiegens der Schichten).

[3] To those who urged him to publish his hypothesis of the double fold, Escher replied invariably: "Nobody would believe me, they would put me in an asylum". (Heim 1919-22).

un-contested leader of Swiss geology and in 1878 published in his book *Mechanismus der Gebirgsbildung*, the theory of the double fold of his teacher. In his objective to reduce to the maximum the impact of the discontinuous faulted overlap, Escher had imagined the idea of two recumbent folds, of opposite vergence, and whose anticlinal hinges, now eroded, would come almost to touch each other at the heart of a pass hollowed out by erosion (Fig. 22a). One such hypothesis (it was his goal) reduces the discontinuous overlay by a half which nevertheless remained close to fifteen kilometres. It is little to say however that already many would find this distance totally improbable.

We will come back to this natural example later, but it is time, at this stage of our story, to turn towards a man who played an indirect but determining role in the discovery of nappes and major overthrust faults; the theorist Eduard Suess. This man from Vienna is the emblematic figure of geology of this period and his work radically influenced the thoughts and ideas of the majority of his contemporaries. Eduard Suess was not a field geologist, and if he liked to compare his ideas to field evidence, his solitary thoughts were elaborated in the secrecy of his own office. In 1875, he published a small book comprising 168 pages which immediately became a classic of geological literature; *Die Entstehung der Alpen*. Simply, he explained that the formation of mountain chains is associated with major *horizontal displacements* resulting from lateral compression. The displacements are all *in the same sense* and the internal compressed zone advances towards and overthrusts the foreland of the orogenic zone. In particular, in the Swiss and Austrian Alps, he recognised that the advance came from the South and that all the movements must be headed for the North. Also, he introduced the concept of periodicity of the tectonic events at the surface of the Earth, insisting on cycles of major transgressions and the sedimentary infilling of geosynclinal basins, then on their subsequent uplift into mountain chains under the effect of compressive movements. This book opened new perspectives to those concerned with the study of tectonics. And numerous will be those who will benefit from this fundamental lesson. Then, from 1883 to 1909, he wrote the three volumes of the famous *Das Antlitz der Erde*. Again, the same ideas are mixed and superbly presented.

The teaching of the master from Vienna was not lost. In 1884, two extraordinary publications precipitated events; the one of the Frenchman Marcel Bertrand and the one of the Scotsmen Peach and Horne.

The reader will have understood from the wording of Lapworth and Callaway's work that the structure of the Scottish mountains lent itself well to the discovery of the new concept. In a short article to the British journal *Nature*, Peach and Horne demonstrated, with arguments based on highly clear and detailed field evidence that a part of the Scottish mountains was overthrusted along a horizontal thrust plane over a distance greater than sixteen kilometres (Figure 23).

Figure 23 - The Moine Thrust in Scotland (modified after Peach and Horne 1884).

This structural discontinuity is described as a fault zone of intense shearing, highly deformed, with the formation of a foliation parallel to the fault plane. Equally, a stretching lineation is described and its orientation gives, in the mind of the authors, the direction of the displacement of the thrusted units. Let's listen to some extracts of their description of the fault contact to appreciate and take into account the degree to which the two Scottish geologists master perfectly the internal strain associated with the displacement of the Moine thrust;

"The intercalation of quartzite are marked likewise by the same streaked appearance, their component particles of quartz and feldspar being all elongated in one common direction. The gneiss associated with the schist above the thrust plane, has had a new set of schistose planes superimposed in it which are on the whole parallel with the thrust plane. All these new structures,...., were obviously connected with the production of the great thrust plane, with which they lie more or less parallel. They point to enormous mechanical movements under which, as the rock sheared, the individual particles were forced over each other in one common direction, viz from east-south-east to west-north-west. Further evidence of this mechanical movement is supplied by certain abundant fine parallel lines, like those of slickensides, which occur almost everywhere on the foliation surfaces or other parallel division planes. These lines run in the same general direction already mentioned." (Peach and Horne 1884, p. 34-35).

1884! What a lesson in the internal strain associated with a thrust fault! One sees here the progress which the Scottish school had achieved at this time in the analysis and understanding of microstructures [4]. But these are not remarkable descriptions which could, in themselves, convince or at least reduce to silence the

---

[4] To better convince oneself of it, it is enough to refer to the Text Book of Geology of Sir Archibald Geikie to which Callaway made reference in the conclusion to his article in 1883. The effects of the internal strain on the appearance of rocks are already described and understood in an amazingly modern way.

opponents of the concept of overthrusting. More was necessary. And this more, in Britain, was the support carried by the dominant figure of Anglo-Saxon geology of the period; Sir Archibald Geikie. By way of introduction to the note in *Nature* Geikie justified the new work in Scotland (of which he said to be the initiator) and asserted, that after re-examination of the Moine region studied in Scotland and in the light of the new descriptions from Peach and Horne, that he no longer stick with Murchison's concept of "anti-nappe"(who never succeeded in taking advantage of Escher's work despite his visit to Switzerland). This conversion of a person of such great authority was enough to stifle all controversies and to throw the Anglo-Saxon geologists along the tracks of a new tectonics.

Let's say, immediately, Marcel Bertrand didn't have this luck. His paper was not in any way like the one of the Scottish geologists. Marcel Bertrand was the geologist who best understood and assimilated Eduard Suess's ideas. His enthusiasm and his admiration for the work of the Austrian man are besides legendary [5]. In particular, the presentation of *Die Entstehung der Alpen* led him to consider mountain chains in a totally new light. And it is like this, in 1884, that he was able to publish an article on the double fold of Escher and Heim *without ever having visited the Swiss Alps*. So it is not necessarily a question of presenting new field data but simply to reinterpret these data (which are otherwise perfect) by the light of the teachings of Eduard Suess. Because what guides the brilliant intuition of Marcel Bertrand are the two important ideas of Suess: firstly the one of major horizontal displacements in mountain chains, and secondly the one of a general movement towards the foreland which implies that the displacements are all in the same sense. For Escher and Heim's double fold, Marcel Bertrand substituted an enormous single overthrust, towards the north, which means towards the Alpine foreland (Fig. 22b). He equally emphasises the analogies which he reveals (or which he believes to discover) with the large overthrust faults demonstrated by Gosselet in the French-Belgian Carboniferous coal-bearing basin, where there the displacements are also all towards the North. Then, he introduces the notion of *zones of slipping* which he defined as zones of ductile deformation, and thus it is like this that he presented the fault contact at the base of the thrust:

"This hypothesis of flow under pressure of deep masses can be visualised in a more precise manner through the hypothesis of the multiplication at depth of

---

[5] Thus, about the first volume of *Das Antlitz der Erde*, Bertrand will say that it "marks a considerable progress, almost the beginning of a new phase, in the study of the major problems in geology" (cited by Termier 1922, p.155) Later, in a lyrical addition, he will say that this book "has marked in the history of geology the end of the first day, the one where the light was" (preface of the first volume of *Das Antlitz der Erde* translated into French by Emmanuel de Margerie).

sliding zones, that is to say again by the possibility for the rock particles to displace themselves slightly each grain in relation to the others; it is this which constitutes plasticity" (Bertrand 1884, p. 324).

At last, he ventures to present a map on which he has represented, on the scale of the whole of Switzerland, the extension of the phenomenon of overthrusts such that they are exposed to the simple examination of a geological map.

To say that this paper went unnoticed would be inaccurate, but it has been welcomed at best with some scepticism and generally with some disbelief. It is necessary firstly to emphasise that geologists have never much appreciated people who interpret the geology of a region without ever having put their feet there. This is not very serious and only some personalities equipped with an exceptional intuition such as Bertrand for the Glarus nappe (overthrust), or later Lugeon for the Carpathes (1903), successfully dared to do this perilous exercise. Also, Marcel Bertrand, who had not yet acquired the renown which he acquired subsequently, did not carry the support of a reputed personality as was the case for the note of Peach and Horne to the English journal *Nature*. One result of this on the European continent, the battle to demonstrate the existence of overthrusts was only beginning. And it is clear that to win this battle, it is above all the Alpine geological community, by far the most numerous and the most influential, that it will be necessary to convince.

In the following years, other field workers demonstrate the extent of overthrusts in mountain chains. In Canada, in the Rockies, McConnel in 1886 described a spectacular thrust which now carries his name. Marcel Bertrand, again, tackles the structure of the Alps in Provence (France) and shows the existence of kilometre long overthrust faults associated with some complete overturning of sedimentary sequences (Bertrand 1887, 1889) [6]. And especially, the Norwegian Törnebohm rewrote the history of the Scandinavian chain by showing that all the mountains have overthrusted towards the south-east (Törnebohm 1888, 1896). The displacements described by the great Norwegian geologist reach values that nobody could have ever been able to imagine. More than one hundred kilometres! To convince others, one most often needs to find

---

[6] In 1887, Marcel Bertrand continued to hammer his certainties strongly: "Thus, in the oldest folding, as in those whose date is closer to us, the same facts are reproduced, and everywhere with an amplification of nature to elude all the predictions...If it is true that one can still discuss the mechanism of these huge overlap phenomenons, neither the existence nor even the generality can be doubted: sooner or later one will be carried to find in these phenomenons the explanation of the "klippes" wrongly attributed to local unconformities, and from now on one cannot refuse to see in them a normal phase of large orogenic movements".

oneself in the heart of the influential circles. And who in the influential circles was concerned about such ideas, from faraway Canada or the mysterious snowy summits of Scandinavia!

So, in this way, in the Alps, in spite of this new work, nothing has changed. In 1891, Heim published a large paper on the Glarus region. Bertrand's hypothesis is not even mentioned. Likewise, in 1893, Sir Archibald Geikie published the third edition of his *Text Book of Geology* where only Heim's hypothesis of the double fold is presented and where, again, Bertrand's hypothesis is not reported (although his work on Provence is mentioned). Yet, was Geikie not particularly well-placed to understand Bertrand's hypothesis? That is to forget that a code of good conduct also exists between the people who dominate a scientific discipline. It is very unusual to argue against the work of a peer of international renown, and for the moment it is out of the question to voice the slightest doubt about Heim's hypothesis of the double fold.

However, one suspects that Bertrand's hypothesis has seduced a very attentive observer of the field discoveries. Already, in 1883, Eduard Suess, in the first volume of *Das Antlitz der Erde*, omitted to speak of the Glaris double fold. This hypothesis of a double overfolding which goes against his theories on the formation of mountain chains can only put him in an awkward position. In 1892, he himself visited the Glaris region and was able to support his ideas such as he had since the beginning. On his return, he goes to Zurich, to meet Albert Heim, and defended Marcel Bertrand's hypothesis. But when Albert Heim asked him if he will publish the idea of a single large scale overthrust towards the North, he replies: *"No, I shall not. YOU must do it yourself, if you can agree with the idea"* (Bailey 1935, p.54). It is easy to imagine the shock that Albert Heim received during the course of this interview.

Time started to play in favour of the new theory. In 1893, Hans Schardt published an article which revolutionised the interpretation of the Pre-Alps from top to bottom [7]. Briefly, in this first article and in a much more detailed way in a subsequent paper in 1898, Schardt proposed the allochthon of the whole of the Pre-Alps, a general movement towards the north and a translation which reached hundreds of kilometres. The stir provoked by these articles is significant. Contrary to Marcel Bertrand's hypothesis on the Glaris Nappe (1884), Hans Schardt supports his work with his own field observations, particularly precise and detailed, by integrating the data from his Swiss colleagues, in particular those of the stratigrapher Renevier. Thus, for the first time, the controversy was establishing itself even in the heart of the Swiss geological community, and the virulent criticisms which Schardt was having to support hurried along a process

---

[7] Read Masson (1976) for a detailed historical overview of Schardt's discoveries.

which was carrying the new theory into the front of the arena, which Marcel Bertrand's article, too rapidly ignored and even scorned by the Alpine geologists, had not been able to do. We will come back to the work of Schardt which is, in many respects, very original in relation to the tectonic ideas of his time. Let's be happy to say for the moment that the discovery of the allochthon of the Pre-Alps by Schardt, by amplifying for a first time the controversy, announced the final act of a discovery of which Escher von der Linth had posed the first milestones more than fifty years beforehand. However, nine years will elapse before this new theory is openly accepted. During this interval, the stormy debates and the acrimonious exchanges were numerous but it is exactly this change of tone in the opponents which indicates, on the contrary, the anxieties of the conservatives and so therefore the progress of the new theory in the geological community.

It is a young Swiss geologist Maurice Lugeon, who succeeded in making the mass of hesitants tip over onto the side of the followers of the concept of overthrusting following a memorable lecture given to the Geological Society of France in Paris on the 1st February 1902. It is difficult to be specific about the reasons which make this particular conference the date of the general conversion of geologists to the new concept, or to explain the great achievement which existed that day for this young geologist of thirty two years. Indeed, his work which embraced all the tectonics of the Pre-Alps and the major part of the Swiss Alps were and still remain a monument in the history of geology which marks a total reinterpretation of the Swiss Alps, still the basis of our current conceptions of this mountain chain. But his brilliant and enthusiastic personality, and skills in argument sufficient to convince even his critics, equally contribute a lot to his succeess. One can also stress that his predecessors, who met more opposition than acclaim, had largely opened the way into which he rushed with enthusiasm, and that also the concept had come to maturation, in particular after the work of Schardt. Finally, one cannot fail to note either that Lugeon's paper, which appeared in the Bulletin of the Geological Society of France in 1902, is preceded by a letter from the uncontested master of Swiss geology, Albert Heim. In this letter, the old Swiss geologist, in admirable terms, congratulated Lugeon for this new synthesis of the Alps and acknowledged that it clarified and explained in a new light all the facts which had remained for him, during such a long time, completely enigmatic. *"That is a real personal joy to recognise that my students go further than me and teach me to accept ideas before which I had until the present stopped"* he adds with a humility which honours a man who through his teaching educated several generations of geologists. This conversion late in life of a master of such authority certainly played an important role, comparable to the conversion of Sir Archibald Geikie eighteen years beforehand at the time of the discovery of the overthrusts in Scotland.

The scientific work of Marcel Lugeon is very fruitful. Until the end of his long life (1870-1953), he published near to three hundred papers of geology, but also numerous studies of dams, water reservoirs, and the formation of the Alpine relief. At the beginning of the century, when Lugeon is already a Professor at the University of Lausanne, two papers (1901 and 1902) each exceeding a hundred pages throw new bases on the Alpine geology. At the reading of the first paper, that is to say before the success of 1902, Eduard Suess wrote to Lugeon, "It will be a long time until you are understood, but I, I have understood you" (cited by Fallot 1954, p.309). The old theorist was mistaken, a year later the theory of overthrusts would triumph.

1902, it is the end of the heroic period. End of the period where a handful of enlightened geologists, convinced of the poverty of the ancient existing discussions obstinately insist on presenting, against the advice of the greatest authorities of the period, a theory and a new conception of the mountain chains under the sarcasm or the indifference of a greater number. Of course, there will be still some resistance and some struggles to impose everywhere the vision of the Alps of Suess, Bertrand, Schardt and Lugeon. We could also present the works of Pierre Termier (1903), Emile Argand (1916), Rudolph Staub (1924) and many others. We could equally turn to the discovery of the nappes in other mountain chains such as the Pyrenees (Bresson 1903, Dixon 1908) or in Belgium (Fourmarier 1923). But the history strictly said of this major discovery breaks off logically at the dawn of the twentieth century when the holders of the new theory have become the majority.

Many names have been cited during the course of this historical overview and one can amaze oneself of the wealth of this period where suddenly so many geologists of such great merit appeared. And yet, one can be certain that many authors are missing, who would have deserved to be mentioned. It is the failing or the advantage of an historical overview to carry out a sorting out which in certain ways has some arbitrary sides. But it is exactly the appearance of all these talented scientists which signifies a major period in the history of tectonics. As in all the scientific disciplines, a period which demands an increased creativity on the part of the scientists always allows the talented to reveal themselves. It is the context of the moment, of an extraordinary potentiality, which gives rise to the maximum expression of each. For example it is the same situation of the constitution of quantum physics at the beginning of the twentieth century with the appearance of all the great physicists such as Plank, Einstein, Bohr, Schrodinger, Heisenberg or De Broglie. It is probable that in a more quiet period (for example after the second world war and during the 1950's), even a Lugeon would not have been able to take the eminent place that he now held in the history of geology.

There are also many injustices in this history because it projects onto the front of the stage some people whilst some giants such as the Norwegian

Törnebohm are in retreat in the background in relation to the place that their immense talent should have given them. These injustices result from the fact that the development of the sciences often possesses a centre of gravity which marginalises those who are situated too far from the spheres of influence. And in this case, the centre of gravity is situated in the heart of the Alpine chain, and in particular in the Swiss Alps. However, the history being written "après coup", certain injustices are sometimes put right. For instance that is the case of Marcel Bertrand who generally occupies the position which he deserves in the historical perspectives dedicated to geology (e.g. Termier 1922, Bailey 1935, Masson 1976, Gohau 1987).

In summary, it is useful to pull some conclusions from this historical overview of the discovery of nappes:

1. In geology, as in the other scientific disciplines (and in all human activities), it is not enough to be right in order to convince others or to be recognised by one's peers. Other external factors associated with the historical context, with the development of the discipline, with the influence of certain influential personalities, and to the degree of the insufficiency of existing theories, have a direct influence on the future of a new theory. Regarding this remark one can recall that the theory of *continental drift* from Wegener was mercilessly remote for the geophysicists to reappear suddenly fifty years later under the modern form of *plate tectonics* thanks to the active influence of these same geophysicists... In science, the "truth" has not in itself any power of conviction.

2. Very often, when a new theory attempts to assert itself, the majority of scientists are deeply hesitant and take position with difficulty. Their wait and see policy favours the existing system. Numerous scientists rely on and support themselves with the attitude of those who already have a well established reputation.

3. It is because of this second point that in Great Britain, the battle ends in 1884 thanks to the conversion of Geikie while on the continent, it needs wait until 1902 (until the conversion of Heim) to see the new theory be widely accepted.

And finally it would be up to Albert Heim to bring this period permanently to an end in his memoire *Geologie der Schweiz* (1919-22) by drawing the Glarus nappe according to the conception of Marcel Bertrand and by citing under the figure *Bertrand 1884*, but also *Suess 1892* as a record of a conversation held at the time when all was still uncertain.

2.2 The Question of Emplacement Mechanisms

Essentially it is for two reasons that so many years were necessary to accept the theory of overthrusts. The first, as we have already indicated, is above all of a psychological nature. It was difficult to accept a theory which completely challenged what was known, or what was believed to be known, about the subject of the structure of mountain chains. The second reason is much more serious because it relates to a scientific argument that even the holders of the new concept were not able to contest or disprove. This argument was developed, not by the geologists, but by the researchers in the field of soil and rock mechanics who were pointing to the unplausible aspect of the postulated horizontal displacements, estimated to be of the order of a hundred kilometres by certain pioneers of the discovery (e.g. Törnebohm 1888). This argument consists of refusing the concept of a nappe, not on the basis of stratigraphic and paleontological arguments developed by the promoters of the new concept, but on the basis of the *mechanical impossibility* of producing such displacements. For example, let's listen to what a reputed researcher of rock mechanics said at that time:

"Have the authors considered that this means movement of a solid block of rock or rocks of unknown length and thickness of 100 miles over the underlying complex of new rocks?... I venture to think that no force applied in any of the mechanical ways known to us in Nature would move such a mass" (Reade 1908).

That is the major question. The problem posed by these overthrusts is the one of the nature of the forces capable of displacing such considerable rock volumes over great distances. But the knowledge of the period about the nature of the forces put into play during the orogenies does not allow a decisive hypothesis to be proposed explaining the driving force of the observed displacements. And when, amongst geologists, the initial resistances are permanently overcome in 1902, the question of the cause and the mechanism of these thrusts is still not resolved. From this time, the history of the nappes is confused with the problem of their emplacement mechanism.

At the outset, two antagonistic mechanisms for the emplacement of nappes are proposed (a) gliding under the influence of gravity along a basal slope, inclined in the direction of the displacement and (b) translation under the action of horizontal forces, and therefore under the action of tectonic forces, applied to the rear of the rock material and which push it along a more or less horizontal surface (Fig. 24). These two mechanisms are centred around the respective roles played by volume forces (forces produced by the Earth's gravitational field) and surface forces (horizontal forces associated with the tectonic episodes). Fairly quickly, since the formulation of the new concept, the modellers of rock mechanics looked into the new problem and attempted to show the mechanical implications of these two major mechanisms. We will analyse their solutions in detail in the following chapter. For the moment it is sufficient for us to say that for almost a century the mechanical modellers, following the example of the geologists in acknowledging overthrusts, have proposed and defended some solutions sometimes diametrically opposite to each other and that their work has equally given rise to tough controversies.

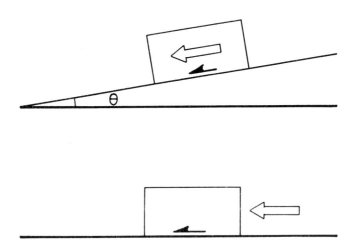

Figure 24 - The two antagonistic mechanisms of the emplacement of nappes proposed since the discovery of the concept. Above: gliding along an inclined plane, below: push from the rear on a horizontal surface.

From the side of the geologists, the principal argument in favour of one or the other of these two models could be only indirect and almost intuitive. At this time the analysis of the internal strain is not sufficiently developed to supply the arguments to resolve this issue which we will review in the fourth chapter of this book. For the majority of the time, the geologists have looked for an explanation of overthrusts in the theories of the formation of mountain chains, and the mechanism fashionable at the time became dependent on the evolution of the ideas in this field of study.

Schematically, the theories on the formation of mountain chains can be categorised together into two groups: (a) the *mobilist* theories which stress the horizontal movements of the superficial outer crust of the Earth and (b) the *fixist* theories which, in reverse, insist on the dominance of vertical movements recorded by the Earth's crust. We have understood that the mobilist theories give some arguments to the holders of the emplacement by compression, while the verticalist theories reinforce the side of those in favour of overthrusts under the influence of gravity. Thus, the geologists working on the nappes are very often enclosed within what is a false discussion and have, in the majority of cases, followed the development of the orogenic theories to argue what was supposed to be THE model of the emplacement of nappes.

From this point of view, since the discovery of overthrusts, three major periods can be distinguished where the 'in fashion' emplacement mechanism appeared to be a simple consequence of the dominant orogenic theory.

From the discovery of nappes (1884) until 1930. Two dominant orogenic theories successively dominate this period. The first was copiously developed by Eduard Suess, following the first works of Elie de Beaumont (1829) and of Dana (1847). At this time, it is considered that the Earth is undergoing permanent cooling and that it contracts regularly under the effect of this loss of heat. This contraction causes large scale vertical subsidence which explains the marine transgressions and the infilling of geosynclinal basins. At the same time, this contraction also causes horizontal movements on the Earth's crust, which becomes too large in relation to the overall contraction. The crust responds by rippling and folding, like the skin of an old apple, in this way giving rise to the orogenies and the uplift of the mountain chains [8]. For the mainpart of the geologists who defend the new concept, it is these horizontal movements and the associated compressions which are the cause of the overthrusting.

It is not amazing that the geologists have attempted from this time in the goal to better convince their contemporaries, to reproduce in the laboratory the major overthrusts observed in nature, by using laboratory-scale reduced models. The most famous of these attempts is the one of the British geologist Henry Cadell (1887), who shared in the discovery of the Scottish nappes with Peach and Horne. Being inspired in part by the experiments of Favre (1878), Cadell prepared alternate horizontal beds of plaster and of wet sand which he compressed laterally with the help of a hand operated piston, or helped by a lever when the mechanical resistance of the beds was too great. By laterally advancing a moveable piston he obtained a stacking of folds and thrusts which, in a certain way, reproduced the overthrusts that the British geologists had analysed in Scotland. The unsophisticated side of these experiments can tempt a smile, but they illustrate that the research of the Anglo-Saxon geologists, well-ahead of the continental

---

[8] This theory will be abandoned when the physicists showed the role played by the breakdown of radioactive elements in the thermal field of the Earth. However, there still remain some supporters of this theory (e.g. Meyerhoff et al., 1992).

European geologists in the acceptance of the new concept (cf historical overview of the discovery), had developed to better understand the orogenic processes which cause the overthrusts. For them, as with all their contemporaries, filled with the ideas of Suess, it is a horizontal compressive push which must explain the formation of overthrusts.

Some years later, the American geologists, well informed about the research and the discoveries of their British cousins, similarly produced scaled models in the laboratory by using an experimental set-up very similar to the one devised by Cadell (Willis 1893, Chamberlin and Miller 1918). There again, it is by a horizontal compression at the rear of the analogue materials that the Americans attempt to explain the formation of nappes and of overthrusts.

It should not be believed, however, that the emplacement mechanism of nappes under the action of gravity be totally overshadowed at this period. In effect, unlike Bertrand, Lugeon or Termier, since his first publications on the Pre-Alps allochthon Hans Schardt supported a gravity model to explain the displacement of these nappes. Through recognising that the creation of the relief responsible for the gravity sliding is due to the contraction of the internal zone of the chain, he pointed out a certain number of geometric elements, such as the presence of a more ductile layer at the base of the thrusted units or the presence of chaotic complexes under the nappes (i.e. olistostromes, cf p. 17) which he interpreted, and moreover correctly in the case of the Pre-Alps, as an indication of a general sliding of the Pre-Alps from a high region towards a low region (read Masson (1976) for more details on Schardt's theories).

Likewise, other authors somewhat at the fringe from the tectonic ideas of the period have equally attempted experiments where the emplacement and the displacement take place under the unique action of gravity. In particular, the experiments of Reyer (1888) where the sliding is achieved following the injection of liquid at the base of the model is a brilliant anticipation of the discovery of the role of fluid pressure in soil mechanics (Terzaghi 1950, cf chapter on the mechanics of thrusts). More modern again, in the light of the cross-sections by Marcel Bertrand, the experiments of Sollas (1906) where folded beds in the Alps are considered to be the result of a viscous flow of the material and no longer as the result of a buckling of beds following upon a compression. The experiments produced are also, therefore, an amazing anticipation of the tectonic flowing of the French-Swiss geologists of the years 1935-1955 or, even better, of the concept of gravitational spreading of which we will speak again in detail in the third chapter. But these contributions on the gravity aspect of overthrusts are very much in the minority.

Later, after the decline of the theory of the contraction of the Earth, when Wegener (1912) proposes the extraordinary theory of continental drift, the majority of geologists will exploit this mobilist theory as new proof of the emplacement of nappes by compression. The nappes are the consequence of major horizontal displacements of the continents on the surface of the Earth, the forces put into play are gigantic and explain the observed overthrusting. From this point of view, it is the Swiss geologist Argand who will succeed in using to the best the Wegener's theory by unifying in a sound way the tectonic data on the mountain chains and the supposed drift of the continents (Argand 1916, 1924).

From 1930 to the 1960's. In this period Wegener's theory unfortunately cannot assert itself because of an almost general opposition from the geophysicists. The field is free for new theories, extremely varied in nature, but which are all, perhaps as a consequence of the relative failure of Wegener, of the verticalist or fixist type theories (Haarman 1930, Van Bemmelen 1933, Beloussov 1962). The most fashionable, following the hypothesis of convection currents in the interior of the Earth (Holmes 1933), is the one of suction and deformation of the geosynclinal basin between two convection cells (Griggs 1939, Umbgrove 1948). This period corresponds to the triumph of the gravity theories under the influence of the French and Swiss schools (Grenoble and Lausanne respectively), and the Italian schools. The nappes flow through the effect of gravity on the limb of a crustal upwhelling (or geotumor) resulting from vertical movements recorded by the geosyncline (Schneegans 1938, Lugeon and Gagnebin 1941, Dal Piaz 1942, Lugeon 1943, Aubert 1945, Trevisan 1946, Gignoux 1948, Moret 1950). The driving role of gravity in the displacement of the nappes is the consequence of the vertical movements [9].

The "flow by gravity tectonics", and its procession of implications on the behaviour of rocks during the course of deformation, became one of the principal axes of research during this period. By retaking (and without suspecting) some of Sollas's ideas, some stimulating analogies are made with the kinematics of

---

[9] It should be noted that some geologists, such as Lugeon, until then supporters of the models of compression become the unconditional holders of the role played by gravity. Such conversions, according to the theoretical fashions, could lend a smile and pose questions about the foundations and the firmness of scientific concepts. One would be wrong. For Lugeon, for example, these conceptual evolutions, determined from the outside through the historical evolution of geology, fully allow him to take part in the theoretical renewal of the problem of the emplacement of nappes, and during this period he presents the definition of new important concepts such as the concept of diverticulation. And if it is true that each science evolves always within a global paradigm, historically destined to be replaced (sooner or later!) by another better paradigm (Kuhn 1962), it is because the new paradigm renews the field of investigation of the discipline and allows obscure points to be clarified, previously inexplicable in the framework of the ancient paradigm. By way of example, the verticalist theories, very much scorned nowadays, have allowed a formidable advance in the knowledge of the emplacement of nappes.

glaciers. The flow of certain gravitational nappes is even compared with the slow movement of glaciers; and equally the influence of the factor of time on the rheology of rocks is glimpsed at by the geological community (Chamberlin 1928, Maillet and Pavans de Ceccaty 1937, Gignoux 1948). Some of Schard's ideas forgotten since, are "rediscovered" and modernised through various field studies. Numerous concepts of the emplacement of nappes by gravity which are concepts still used in the present by the majority of field geologists, date back to this period.

Plate tectonics of the present day. Towards the end of the 1960's, the earth sciences experience one of the most significant epistemological ruptures of their history. The new geophysical data on the age and on the paleomagnetism of the oceanic crust are suddenly put together by several authors, who elaborate with an amazing rapidity, undoubtedly due to the legacy left by Wegener, the theory of plate tectonics. The fruitfulness of this new global model is such that it is adopted almost immediately, and with enthusiasm, by the entire geological community. For the nappes, it is the decline of the gravity theories because the plate tectonic mobilist theory replaces Argand's ideas on the formation of mountain chains back into the arena [10].

In parallel, spectacular progress is made in the knowledge and the qualitative and quantitative analysis of internal strain (Ramsay 1967). We will speak about that at length in the chapter devoted to the kinematics. Let's note only that this progress has a significant impact on the gravity-compression alternative, because it links up with another argument from the previous period, an argument which was regarding this issue of the gravity-schistosity alternative. For the authors of the preceding period, the absence of internal strain, and therefore of schistosity, is a determining factor to highlight the driving role of gravity. It is effectively an argument, but does it demonstrate that gravity and schistosity are incompatible? Besides, this argument, developed by the supporters of the role of gravity, turns against them during the following third period. The recent analytical studies on the internal strain of rocks help to support the rejection of the hypothesis of the emplacement of nappes through gravity. In effect, it can be shown that certain classic examples of gravitational nappes, considered up until that time to be unharmed of internal strain, are in reality affected by an internal strain which is far from negligible. In fact, this problem issue is not one of them

---

[10] If the prophetic reconstructions of Argand are so much respected by the current geologists, it is mainly because the conceptions of today's geologists are deeply rooted in the paradigm of plate tectonics. This respect illustrates perfectly the fact that "one lends to former scientists all the more credit that their theories are close to our current conceptions" (Gould 1988, p. 243).

and must be challenged. In a purely theoretical perspective, there is nothing in the theory of deformation which suggests that gravity and internal strain are mutually exclusive. Schistosity develops in a material when it deforms in a ductile manner. The conditions of ductile deformation are unrelated either to the types of forces (of volume or of surface), nor to the overall mechanism of deformation (compression, distension...). This false problem, inherited from ancient arguments, has now been superseded, but it does reappear suddenly sometimes, sporadically, because it responds to an intuitive (but false) perception of the reality; the difficulty in understanding that rocks (apparently so hard!) may deform under the sole action of gravity (but at a scale of millions of years).

Nowadays, these scientific quarrels on the compression-gravity alternative have become insignificant because no serious geologist would contemplate denying either the possibility of emplacement through compression, or the possibility of emplacement under the influence of gravitational forces. Besides, one has witnessed, since the end of the 1970's, a stimulating struggle between two theories which can eventually be combined. One is purely gravity, it is the theory of *spreading* (Price 1973, Elliot 1976a), borrowed from the glaciologists (Nye 1952), which sweeps aside many objections made generally to the emplacement mechanism through gravity, in particular the eternal problem of the basal slope (cf Lemoine 1973). The second, resolutely compressive, is the one of the *critical wedge* which eliminates the usual objection concerning the impossibility of emplacement of volumes of rock material as a complete block whose surface is too great in relation to their thickness (Chapple 1978). These two models and the rest of the research into the area of rock mechanics are developed in the following chapter.

CHAPTER 3

3 MECHANICS

3.1 The Mechanical Paradox

In 1959, two Americans, Hubbert and Rubey, published a paper which immediately became a classic of the geological literature and aroused a substantial renewal of interest in the mechanical studies on the emplacement of nappes. On the tracks of the Austrian Smolukowsky (1909), but making use of the modern knowledge of rock mechanics, Hubbert and Rubey defined the problem of the emplacement mechanisms of nappes from a purely mechanical perspective, qualifying this problem as a mechanical paradox, and finally proposing a solution which aimed to be all embracing. Let us look immediately at the main part of the mechanical argument of this famous paradox like it is presented at the time by the two Americans.

Maximum length of a nappe. The first mechanical analyses of the emplacement of nappes considered the displacement of a non-deformable rectangular prism on a horizontal surface (Fig. 25). If a sufficient force F is applied to the rear of the prism, the rupture which occurs at the base of the prism is governed by the Navier-Coulomb criteria of failure (brittle deformation). In this case, the normal stress (σ) and the tangential stress (τ) along the fault plane are linked by the equation:

$$\tau = \tau_0 + \sigma \tan \phi \qquad (1)$$

where τ_0 and ϕ are two constants associated with the rock types and are called respectively the cohesion and the angle of internal friction, the second term being more generally referred to as the coefficient of internal friction μ ($\mu = \tan \phi$).

On the other hand, in two-dimensional experiments produced in the laboratory, the two principal stresses (σ_1 and σ_3) are linked by the following equation when failure occurs:

$$\sigma_1 = a + b\,\sigma_3 \tag{2}$$

where:

$$a = 2\tau_0\sqrt{b} \tag{3}$$

$$b = \frac{1 + \sin\phi}{1 - \sin\phi} \tag{4}$$

If the rectangular prism is of dimension x, y and z (Fig. 25), the resistance to the movement under the influence of force F is equal to:

$$F = (xyz)\,w\mu \qquad \text{(Smolukowsky, 1909)} \tag{5}$$

with w as the volume mass of the rock.

This equation can be written as a function of the normal horizontal stress σ_{xx} applied on the surface yz ($\sigma_{xx} = F/yz$) in the direction of the axis x:

$$F = \sigma_{xx}\,zy \tag{6}$$

Since acceleration due to gravity is negligible in geology, the sum of the forces in the direction of x must be equal to zero (Fig. 25). Therefore:

$$\sigma_{xx}\,z = \tau_{zx}\,x \tag{7}$$

Following the Mohr-Coulomb equation (equation 1), and by disregarding the cohesion τ_0, the tangential stress τ_{zx} on the basal plane can be expressed as a function of the normal stress σ_{zz} linked to the weight of the nappe:

$$\tau_{zx} = \mu\,\sigma_{zz} \tag{8}$$

By combining equations (7) and (8):

$$\sigma_{xx} = \frac{\mu\sigma_{zz}x}{z} \tag{9}$$

then by replacing this new value of σ_{xx} in equation (6):

$$F = \sigma_{zz}\,xy\,\mu \tag{10}$$

by rearranging equations (6) and (10), it is possible to obtain the maximum length of a nappe which is displaced as a coherent 'block' under the effect of a horizontal stress applied at the rear:

$$x = \frac{\sigma_{xx}z}{\sigma_{zz}\mu} \qquad \text{(Price and Cosgrove, 1990)} \tag{11}$$

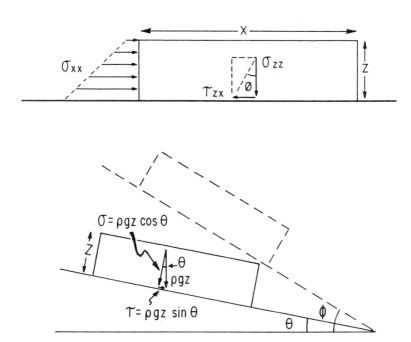

Figure 25 - Initial conditions of the mechanical model of Hubbert and Rubey. Above: displacement by back push of a non-deformable rectangular prism, below: gliding on an inclined plane of a non-deformable rectangular prism.

This maximum length is essentially a function of the thickness (z) of the nappe but is independent of the width (y) of the thrusted unit. The normal stress σ_{zz} at the base of the nappe is equal to the weight of the column of rock (z):

$$\sigma_{zz} = \rho g z \tag{12}$$

where ρ is the density of rock and g is the constant of gravity. Knowing σ_{zz}, the value of σ_{xx} can be calculated, its mean value from the top to the base, by using equation (2). By taking the universally accepted values for the majority of the constants necessary to the calculation ($\mu= 0.577$, $\rho= 2.30 \text{gcm}^{-3}$, $\tau_0 = 20\text{MPa}$), it is stated that a nappe of a thickness of 1 kilometre cannot exceed a length of 8 kilometres and that a nappe of 6 kilometres thickness cannot exceed 21 kilometres in length. This is because the strength of the rock is not infinite; beyond a certain length related to the thickness, the rock mass can not only translate itself rigidly but can deform through folds and reverse faults (Fig. 26). Therefore, to displace a nappe of thirty kilometres in length on a horizontal surface by a force applied at the rear appeared to be mechanically impossible (Smolukowsky 1909, Hubbert and Rubey 1959).

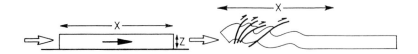

Figure 26 - The mechanical properties of rocks limit a block pushed from the rear to a length X as a function of the thickness Z.

Gliding on an inclined plane. The minimum angle in order to produce a fracture plane along an inclined plane of angle θ under the action of gravity can equally be calculated from the Navier-Coulomb failure criteria. The normal stresses (σ) and tangential stresses (τ) on the fracture surface are given by the relationships (Fig. 25):

$$\sigma = \rho g z \cos \theta \tag{13}$$
$$\tau = \rho g z \sin \theta \tag{14}$$

from which can be extracted:

$$\frac{\tau}{\sigma} = \tan \theta \tag{15}$$

after equation (8), it is known that:

$$\frac{\tau}{\sigma} = \mu \tag{16}$$

μ being close to 0.58 for the majority of sedimentary rocks:
$$\theta = 30° \tag{17}$$

The failure along a sliding plane under the sole action of gravity cannot occur with an angle of less than thirty degrees (Hubbert and Rubey 1959). Besides it will be remarked that such a slope would require absolutely absurd altitudes before the thrusting occurs (Lemoine 1973).

Mechanical paradox or ill-posed problem? This first mechanical approach on the displacement of nappes immediately poses a problem which all the previous works are concerned about: in one respect the nappes are too thin in relation to their surface area to be displaced as a unit block by a force applied to the rear, in another respect the angle observed in the field between the basal surface and the horizontal is too low in order that gravity be the cause of sliding under the actual weight of the nappe itself. This is Hubbert and Rubey's mechanical paradox. It is effectively a paradox because the mechanical analysis seems to reject as much the emplacement by gravity as the emplacement by compression while, at the same time, thrusts are an indisputable geological reality.

However, it is suspected that this famous mechanical paradox hides above all an ill-posed problem. In 1909 the Austrian Smolukowsky was the first to demonstrate the difficulty of transmitting the horizontal stress in the rock material to push it rigidly without rupture and deformation. Yet his conclusion is not a rejection of the so recent and yet so highly contested concept of thrusting. On the contrary, he suggested a solution by insisting on the necessity of taking into account a viscous rheological behaviour, and of accommodating the factor of time, so fundamental in geology:

"...then the law of viscous liquid friction will come into play, instead of the friction of solids; therefore any force, however small, will succeed in moving the block. Its velocity may be small..., but in geology we have plenty of time; there is no hurry" (Smolukowsky 1909, p. 205).

This suggestion has not been followed or understood by Smolukowsky's contemporaries and it was necessary to wait near to sixty years to see tested the hypothesis of a viscous behaviour in the emplacement of nappes (Bucher 1956, Kelhe 1970). Nevertheless, Sollas's experiments could be cited here which at the beginning of the century, were already testing this hypothesis experimentally.

Likewise, as early as 1948, Jean Goguel presented a remarkably modern mechanical analysis. In his paper he analysed different possible models for the emplacement of nappes by integrating a layer of weak mechanical strength at the base (such as a layer rich in gypsum) and by considering an ideally plastic rheology for the rock material. He showed how a nappe of three kilometres thickness can slide on a slope of three degrees, and a nappe of six kilometres on a slope of one and a half degrees. Clearly, as Voight (1976, p.157) points out, the mechanical paradox of large nappes, even before its formulation, no longer existed! But we are here confronted by one of these stammerings of science, where a work went unnoticed, either because it is too ahead of its time, or because it does not have the characteristics to assert itself within the international scientific community. In this case, Goguel's work has certainly suffered by its having been written up in French and so in this way to be almost inaccessible to the Anglo-Saxon scientific community.

But this does not explain everything. At the beginning of their article, Hubbert and Rubey briefly refer to Smolukowsky's suggestion, only to forget it totally afterwards. It is clear that for these two authors a presentation as "paradoxical" of the problem of the emplacement of nappes has the intention of introducing their solution (the role of fluid pressure at the base of nappes) as the only possible. Without wanting to criticise these two authors, in order to give more weight to their own model it is likely that they present the mechanical analysis of nappes under as simplistic a problem. In effect, it is only with difficulty that one can believe that they would ignore all of the field descriptions of the period where frequently the presence of a decollement level of low mechanical strength was clearly established. In fact, the real problem, which is not tackled in Hubbert and Rubey's paper, is the one of the rheology of the rocks at the time of the overthrusting. However it may be, it is from this apparent mechanical paradox that the researchers of rock mechanics have subsequently placed themselves to defend a viable mechanical model of the emplacement of nappes.

3.2   Resolution of the Mechanical Paradox

3.2.1   HYPOTHESIS OF FLUID PRESSURE.

The solution proposed by Hubbert and Rubey to solve the paradox borrowed the discovery of the role of the pressure of fluids in large landslides from the soil mechanics literature (Terzhagi 1945, 1950). It is known that numerous sedimentary rocks are porous and that these pores are generally saturated by water. In this case, the problem no longer relates to the previous analysis where the rocks are considered "dry", because the pore fluid creates a hydrostatic pressure which modifies the Navier-Coulomb fracture criteria. This hydrostatic pressure (P) at depth z is equal to :

$$p = \rho_w \, g \, z \tag{18}$$

where ρ_w is the density of water.

At depth, the vertical stress is related to the weight of the column of rock at the point considered (equation 12):

$$\sigma_{zz} = \rho \, g \, z \tag{12}$$

where ρ is the density of the total rock, where the water is included.

The isotropic pressure of a fluid, by exerting itself (by definition) in all directions, reduces the normal stresses applied to the material (Terzhagi 1945). Thus, at depth, the actual normal vertical stress, called the effective stress, is equal to :

$$\sigma_e = \sigma_{zz} - p \tag{19}$$

On the other hand, the tangential stresses are not modified by the isotropic fluid pressure which always remains normal to the wall of the pores of the rock. By retaking the preceding example of a block of rectangular dimension compressed at the rear, and by taking account of the effective stress while still ignoring the cohesion (τ_0), the Navier-Coulomb fracture criteria is written:

$$\tau_{zx} = (\sigma_{zz} - p) \, \mu \tag{20}$$

It is stated that the increase of fluid pressure reduces the normal stress on the fracture plane but that equally it reduces the tangential stress and therefore the resistance to sliding. In the borderline case where the fluid pressure approaches a value close to the weight of the rock column, the resistance (τ_{zx}) becomes practically negligible. Therefore, taking the fluid pressure into account dramatically changes the data of the problem. By putting $\lambda = p/\sigma_{zz}$, the ratio between the density of water and the density of saturated rock, equation (20) is written under the form:

$$\tau_{zx} = \sigma_{zz} (1-\lambda) \mu \tag{21}$$

Hubbert and Rubey therefore recalculated the length of a nappe pushed on a horizontal surface as a function of the value of this coefficient λ. In particular, where the ratio λ is equal to 0.8, a nappe of five kilometres thickness could measure near to sixty kilometres length and a nappe of eight kilometres thickness could approach eighty kilometres long.

Likewise, in the case of gliding on an inclined plane, the angular condition of the rupture (equation 15 and 16) is modified in the following way:

$$\tan \theta = (1-\lambda) \mu \tag{22}$$

For values λ approaching 0.9, sliding becomes possible for angles which are less than five degrees (Hubbert and Rubey 1959).

In a second paper, Rubey and Hubbert (1959) endeavour to show some natural examples where one can assume that the allochthonous unit has been the place of an elevated fluid pressure at the time of the displacement. The mechanical paradox is considered to be solved. Since its formulation in 1959 the hypothesis of the pressure of fluids knew a considerable interest. And Hubbert and Rubey's paper undoubtedly marks a turning point and a renewal in the mechanical study of large thrusts.

However, from a purely mechanical point of view, their work is far from being exempt of criticism and several authors will apply themselves to demonstrate it. The first criticism concerns the importance of the cohesion of material (τ_0) which is totally neglected in the failure equation (equation 1). For Hubbert and Rubey, a fracture along so large a surface propagates itself like a dislocation so that the cohesion must be taken into account only at the actual place where the fracture propagates itself, which means on a surface so small in relation to the extent of the fault that the cohesion can therefore be neglected. This returns to the consideration that the displacement takes place along a pre-existing fracture. For certain authors, one such simplification is unacceptable (Birch 1961, Hsu 1969). In particular, Hsu (1969) recalculated the maximum length of a nappe pushed at the rear by taking the cohesion into account. He states that in this case the maximum length of a nappe is considerably reduced and concluded that the fluid pressure may not serve to act as a panacea to explain the displacement of nappes. He also insists on the presence of a level which undergoes ductile deformation at the base of numerous nappes and therefore on the necessity of taking into account other deformation mechanisms than the Navier-Coulomb failure criteria.

An even more serious criticism concerns Hubbert and Rubey's analysis of stresses in the rectangular block resting on a horizontal surface. In effect, this analysis contained a tacit and unquestioned hypothesis which proved to be undoubtedly wrong. The principal stresses (σ_1 and σ_3) are considered to be perpendicular to the edges of the prism and are likened to the horizontal stress (σ_{xx}) applied at the rear of the prism and to the vertical stress (σ_{zz}) related to the weight of the nappe. Yet it is clear that the tangential stress (τ_{zx}) parallel to the base of the prism is not zero, that the analysis is therefore not made following the principal planes of the stress ellipse and that equation (2) may not be used since the horizontal stress applied at the rear of the prism just as the vertical stress associated with the weight of the nappe does not relate to the principal stresses of the system studied (Birch 1961, Forristall 1972). This was anyway already known since Hafner (1951). By using the theory of elasticity to calculate the state of stress in the block, and Hubbert himself (1951) in using the Mohr-Coulomb theory, had calculated the principal stress trajectories, just like the ones of the potential reverse faults (Fig. 27). These potential reverse faults have a concave form and present a listric shape just as is often observed in the field. Some reverse faults of this type have even been reproduced experimentally by Hubbert (1951). Equally by using the theory of elasticity to calculate the state of stress in the block, Forristall (1972) shows that the maximum length calculated by Hubbert and Rubey is in general two times greater than the realistic value allowed. Thus he concluded, if one wants to displace a block of large dimension, it is absolutely necessary that there is (a) an abnormally high fluid pressure localised in a level parallel to the base of the nappe and (b) a basal level of low cohesion.

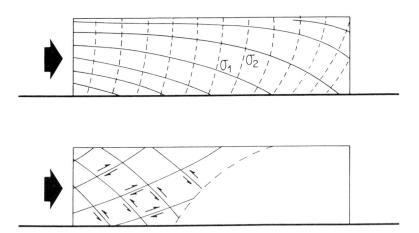

Figure 27 - Trajectories of the principal stresses (above) and potential reverse faults (below) with a horizontal compression at the rear of a prism of rectangular dimension (modified after Hafner 1951).

In spite of these criticisms, there is no doubt that fluid pressure can play a significant role in the formation of nappes and overthrusts [11]. In this respect, numerous researchers have emphasised the importance of the dehydration of certain minerals in the production of water (for example the transformation from gypsum to anhydrite, Laubscher 1961, Heard and Rubey 1966), the ascension of granite within a sedimentary pile (Platt 1961) or in a more general manner the metamorphic reactions which most often are accompanied by a dehydration of hydrous minerals (e.g. Goguel 1969, Ayrton 1980). Equally it can be shown that rauhwackes which very often mark out the contacts of nappes have been interpreted as resulting from the hydraulic fracturing of highly porous dolomitic type rocks (Masson 1972). Some of these examples strongly suggest that the fluid pressure undoubtedly reaches a critical value only at the base of the allochthonous units (i.e. in the decollement zone) and that the actual situation is such that $\lambda_{decollement} > \lambda_{rest\ of\ the\ nappe}$. This leads to the notion of planes of super-weakness at the base of potential nappes and thrusts, a decollement zone where the cohesion would be weak in an environment of high fluid pressure (Gretener 1972, 1981).

3.2.2 HYPOTHESIS OF A LOW VISCOSITY BASAL LAYER.

Contesting that a sufficiently high fluid pressure can exist at the base of nappes observed in the field, Kelhe (1970) makes his hypothesis a combination of both Smolukowsky's idea, apparently without knowing that he was doing so, and also the idea of Hsu (1960) of a displacement achieved by the viscous deformation of a ductile layer situated at the base of the thrusted unit. Rejecting the conception of a clearly individualised basal fault (brittle behaviour or discontinuous deformation), he returned to the notion of a decollement level, which means a return to the notion of a basal zone where an enormous shear strain responsible for the displacement is concentrated while the upper part of the nappe is transported passively and, owing to this fact, can even be unaffected by internal strain. That is the case, for example of the Numidian sandstone of Algeria, displaced without internal strain (or almost without) over nearly one hundred kilometres along a decollement level comprised of Triassic gypsum (Mattauer 1958, p.444 to 469). The problem of the emplacement of nappes is clearly posed in terms of the

---

[11] Some could be amazed that an article containing such a contestable approximation (the omission of cohesion in the calculations), probably a voluntary error in the mechanical analyses (the position of the principal planes of the stress ellipsoid) and which ultimately offers nothing new from the mechanical point of view, since the main part of the analysis is taken from Terzhagi, became so quickly and remained so famous until now. It can be there that the real paradox resides! In reality, these are above all the historical conditions which allow the impact of this article to be explained (cf Voight 1976, p. 261 - 262).

rheology of the material at the time of the overthrust. The solution suggested by Kelhe is close to the one proposed by Goguel (1948) apart from the fact that the basal layer is not analysed as an ideally plastic material but as a Newtonian type viscous material or at least as a non-linear viscous material (power law with a stress exponent). But as he himself points out (Kelhe 1971, p.2684) a highly non-linear viscous material behaves in a way similar to an ideally plastic material, so that the two models are very comparable, if not in their mathematical formulation, at least in the fact that the role played by a basal layer of low mechanical strength is taken into account.

Kelhe analysed in detail the case of sliding along a basal slope inclined at an angle (θ) in relation to the horizontal (Fig. 28). On a small element (Fig. 28), the tangential stress is parallel to the axis x and is a function of the weight of the vertical column of rock ($\rho g y$):

$$\tau_{xy} = -\rho g \sin \theta \, (h-y) \tag{23}$$

where ρ and h are the density and the thickness of the rock material sliding on the slope, g is the gravity constant and y the co-ordinate normal to the slope.

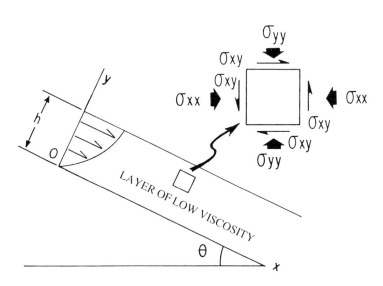

Figure 28 - Heterogeneous simple shear (arrows) along a layer of low viscosity and infinitesimal element used in the analysis of stresses (modified after Kelhe 1970).

In the basal layer where the rheological behaviour is to be considered viscous, the flow law is given by the equation:

$$\tau_{xy} = (\eta/2)\frac{\delta v}{\delta y} \qquad (24)$$

where η is the viscosity of the material and v the velocity rate. By combining the equations (23) and (24) the following is obtained:

$$\frac{\delta v}{\delta y} = (2/\eta)\rho g \sin\theta(h-y) \qquad (25)$$

and by integrating this equation:

$$v = -(\rho g \sin\theta/\eta)(h-y)^2 + C \qquad (26)$$

The value of the constant (C) can be determined for the particular case where the rate is zero, which means on the x-axis (Fig. 29). From where, when y = 0:

$$C = (\rho g \sin\theta/\eta)h^2 \qquad (27)$$

the displacement rate of the nappe (v) is therefore given by the formula:

$$v = (\rho g \sin\theta/\eta)(2hy-y^2) \qquad (28)$$

The velocity of the nappe is estimated at the point y = t (where t is equal to the thickness of the decollement zone, Fig. 29), and can also be written under the form:

$$v = (\rho g \sin\theta/\eta)(2td + t^2) \qquad (29)$$

where d is the thickness of the upper part of the nappe situated above the decollement level (d = h-t) (Fig. 29). This model solves the problem of the basal slope since a slow but definite flow can be achieved on some very low angle slopes of the order of several degrees.

Using a natural example (Carletonville, South Africa) where h = 5km, t = 0.02km and $\theta = 18°$, Fletcher and Gay (1971) show that the displacement rate was of the order of 1.6cm/yr and that the observed displacement of 5 kilometres had reasonably taken place in a bit less than one million years. But it should be noted however that Kelhe's equation is very dependent on the value of viscosity chosen for the decollement zone just as on the thickness of this zone, so that these rates must be considered at best as of the orders of magnitude. On the other hand, it is without doubt more realistic to consider that the rocks have a non-linear viscosity (power law with a stress exponent) and follow a flow law of the type:

$$\dot{e} = A(T)\sigma^n \qquad (30)$$

where $\dot{e}$ is the strain rate, σ the stress, A a constant which is a function of the temperature and n the stress exponent which can vary between 1 (Newtonian-type behaviour) and infinity (in an ideally plastic body). For a marble, n would be close to eight (Heard 1963). By retaking the example of Carletonville and by taking into account a stress exponent close to eight, it can be demonstrated that the displacement would be extremely slow (< 0.1 cm/yr) for a slope less than 10° and

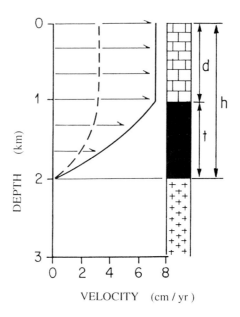

Figure 29 - Vertical velocity profiles (full line: Newtonian viscosity, dashed line: non-Newtonian viscosity) and velocity vectors (arrows) associated with the deformation of the basal layer (in black) (modified after Kelhe 1970).

would become excessively fast (>50cm/yr) where the slope would exceed 25° (Kelhe 1971). To this subject, it should be pointed out that the most recent field studies show that the displacement of shallow nappes is in general rather slow, of the order of 0.1 to 0.2 cm/yr (e.g. Borgia et al. 1992, Mugnier et al. 1992, Suppe et al. 1992, Taboada et al. 1993, Shaw and Suppe 1994).

It should be noted, since we have already spoken about it during the course of the historical perspective of the discovery of nappes, that the Glarus nappe in the Swiss Alps comprises a mylonite zone at its base (the "Lochseiten calc mylonite") and that the displacement of the nappe occurred through the pseudo-viscous deformation of this basal layer overriden by rocks considered to be rigid or to be slightly deformable (Hsu 1969, Schmid 1975). But numerous natural examples can be used to illustrate Kelhe's model. The most famous case is the one of the Bearpaw Mountains in Montana (U.S.A.) where the sedimentary cover sequence has slid along the bentonitic schists from a topographic dome formed due to the underlying intrusion of magmatic rocks (Reeves 1925, Gucwa and Kelhe 1978). Likewise, the ascension and emplacement of the Idaho Batholith (U.S.A.) caused the gravitational gliding towards the East of the overlying sedimentary sequence (Scholten 1973). This model lends itself equally well to be verified experimentally and several authors have succeeded in producing gravitational gliding on a layer of low mechanical strength during laboratory experiments (e.g. Blay et al. 1976, Guterman 1980).

Several researchers have proposed some mechanical models putting the accent on the existence of a basal layer of low mechanical strength. The discussion always turns to the rheology of the basal layer and of the rock unit overriding it. It is necessary to confess that as yet little is known about the rheology of rocks and that for the mechanical studies this uncertainty constitutes sufficient space for some models to be proposed which have varied enough hypotheses on the rheological behaviour of rocks during thrusting. Among these models, the model of Ramback and Deramond (1979) can be noted as an example, where the upper (thick) block has a plastic behaviour (with a very elevated yield stress) while the basal layer follows a visco-plastic behaviour (Bingham's body); and also the model from Mandl and Shippam (1981) can be noted, where the upper block behaves elastically while the basal layer deforms plastically.

3.2.3　HYPOTHESIS OF GRAVITATIONAL SPREADING.

The model of gravitational spreading is by far the one which raises the most criticism amongst the geologists and the researchers of rock mechanics. It originates from the principal that the nappe in its entirety can behave in a viscous manner and flows under its own weight. This model is a simple transposition into geology of the mechanical explanation of the movement of glaciers (Nye 1952). The movement of the nappe is likened to the flow of a glacier. This means that as the material is incompressible, the vertical shortening of the nappe under its own weight (gravitational collapse) is accompanied correspondingly by a horizontal lengthening (gravitational spreading). It is this spreading which strictly speaking corresponds to the overthrust. The interest in this gravity model is that it addresses the problem of the basal slope, so crucial in Hubbert and Rubey's pseudo-paradox. In effect it is the slope of the upper surface, and not a hypothetical basal slope, which determines the movement of the material which is tending towards a gravitational equilibrium, when this material can flow under its own weight. At the limit, the displacement can occur in the reverse sense from the basal slope if the upper surface slope is sufficiently high that it can trigger gravitational instability. Furthermore, it can be seen that this model is not without similarity to the model of tectonic flowing through gravity of the Francophone and Italian authors of the 1940's (cf Roubault 1949, p. 222).

It is the Americans Price and Mountjoy (1971) who first proposed that this model be applied to the Canadian Rockies but it is another American, David Elliott (1976a and b) who succeeded in placing this model of the emplacement of nappes in full view of the geological community by developing in a detailed way all of its mechanical implications. In particular, and it is a troubling point which appears to highlight well the role of gravity, Elliott shows that the quantity of displacement in the Canadian Rockies is directly proportional to the volume (and therefore to the mass) of the thrusted sheets. The greater the volume of the thrust units, the greater is the displacement (Fig. 30).

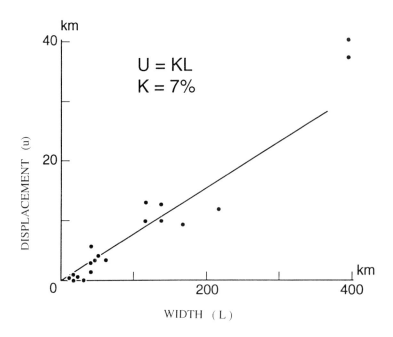

Figure 30 - Displacement (U) of allochthonous units as a function of the lateral extension (L) of the thrusts in the Canadian Rockies. The straight line U=KL shows that displacements are directly proportional to the volumes of the thrusted units (modified after Elliott, 1976b).

The key equation of the mechanical model is simple enough to demonstrate. Following Ramberg (1977), a vertical column of rock of width Δx within a nappe of thickness H which has a small surface slope α is considered (Fig. 31). The deformation occurs as plane strain so that there is no movement in the plane normal to the plane of section. If it is supposed that this occurs under lithostatic conditions ($\sigma_{xx} = \sigma_{yy} = \sigma_{zz}$), and therefore that the normal horizontal stress is equal to the normal vertical stress related to the weight of the nappe:

Mechanics

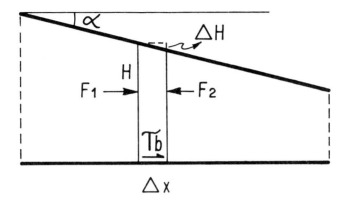

Figure 31 - Initial mechanical conditions of the gravity spreading model (modified after Ramberg, 1977).

$$\sigma_{xx} = \sigma_{yy} = \rho g h \qquad (31)$$

By integrating from 0 to H, the horizontal force (F_1) acting on the left part of the vertical column is equal to:

$$F_1 = \rho g \int_0^H h\, dh = 1/2 \rho g H^2 \qquad (32)$$

while on the right hand side by integrating 0 to (H-ΔH), force F_2 is equal to:

$$F_2 = \rho g \int_0^{H-\Delta H} h\, dh = 1/2 \rho g (H^2 - 2\Delta H H + \Delta H^2) \qquad (33)$$

ΔH^2 being negligible, the force which tends to push the column towards the right in the direction of the upper surface slope is equal to:

$$\Delta F = F_1 - F_2 = \rho g \Delta H H$$
(34)

This force is equal to the tangential stress (τ_b) exerting itself on the small segment Δx:

$$\Delta F = \rho g \Delta H H = \tau_b \Delta x \qquad (35)$$

from where, when ΔH and Δx become infinitesimally small:

$$\tau_b = \rho g H \frac{dH}{dx} = \rho g H \tan \alpha \quad \text{(Goguel 1948, p.404 and Nye 1952)} \quad (36)$$

Thus, it is the slope (α) of the upper surface which determines the sense and the magnitude of the tangential stress at the base of the nappe, and not the dip of the basal slope. In so far as this theory can be applied to a shallow surface nappe of relatively moderate dimension, the authors have stressed the effect that the global topographic slope could have which regularly descends from the high summits found in the heart of mountain chains towards the plains of the foreland. This slope could cause gravitational collapse at the scale of the entire mountain chain. This idea of a gravitational collapse of the chain from high regions towards low regions was proposed for the first time by Bucher in 1956 and in 1962 in a series of experiments on scaled models.

As Ramberg (1977) points out, equation (36) is attractive because of its surprising simplicity but it is incorrect. If the distribution of normal stresses in a nappe which flows under its own weight are considered, it is stated that the true horizontal stress is lower, and undoubtedly considerably lower, than the one corresponding to lithostatic conditions and is directly related to the weight of the nappe (ρ g h). In effect, it is the weight of the nappe which vertically compresses the material and which, at the same time, causes a horizontal extension. In this case, an extensive horizontal deformation associated with a vertical compressive deformation requires a lower normal horizontal stress than the normal vertical stress. Therefore, equation (31) is not valid for σ_{xx} and the shear stress at the base of the nappe is less than the value given by the equation (36). However, as for Hubbert and Rubey's hypothesis of the fluid pressure an incorrect mechanical analysis does not invalidate the model [12]. Simply because, as is suspected, the mechanical analyses are without outcome on the geological phenomenons! Equation (36), even incorrect, has been used since near to fifty years by glaciologists to explain spreading of glaciers, spreading which is in any case indisputable. Nevertheless this equation describes a certain reality because it accommodates the fact that the movement of glaciers (and speaking only of glaciers) is related to their upper surface slope and that the movement stops as soon as the upper surface slope recovers to the horizontal.

The modellers of rock mechanics or the geologists who reject this model (e.g. Mandl 1988, p. 168) essentially contest that rocks can behave like viscous fluids [13]. However it must be emphasised that nothing is known about the rheology of rocks at the scale of a million years and that this factor of time cannot be underestimated. It is perhaps significant that the physical deformation mechanism generally observed in shallow crustal environments is that of pressure solution (e.g. Nickelsen 1972). The flow law of the physical mechanism is given

---

[12] Ramberg is otherwise an unconditional supporter of it.

[13] It is possible that this resistance is above all of a psychological nature as Maurice Gignoux considered it to be as early as 1948: "Basically, what prevents many geologists to come around to the tectonic flowing through gravity, it is their instinctive repugnance to admit that solid rocks can flow like liquids".

by a linear relationship between stress and strain rate (Rutter 1976), that is to say that no yield stress is detectable at the appearance of deformation by pressure solution and that the rheological behaviour is analogous to the deformation of a viscous material. This deformation mechanism is active from low temperatures and excessively slow strain rates (time factor). It should be noted moreover that those who, since the beginning, defended this model have not assumed a particular rheology of the material (cf Elliott 1976). For example for Price (1973), the problem of the rheology is above all a problem of scale: rocks deform in a brittle way at the scale of the outcrop and it is only the deformation on the scale of the thrust belts which can be likened to a viscous type flow. However that may be, the model of gravitational spreading is one of the mechanisms mentioned at the present time to explain (in part) the observed extension in mountain chains at the end of major orogenic cycles.

3.2.4   HYPOTHESIS OF THE TRIANGULAR SHAPE.

A. The critical wedge theory. Abandoning the mechanical analysis of the overthrust of a simple unit, generally represented by a block of rectangular dimension, some authors have investigated the problem of the displacement observed at the scale of thrust belts in the foreland of mountain chains. At this scale, the displacement is no longer analysed in terms of an isolated nappe or ideal thrust sheet but instead an hypothesis is put forward that the entire belt can be analysed as a coherent unit and that the multiple thrusts and imbrications observed within it are only second order phenomenons associated with the deformation and the displacement of the whole of the belt. Credit must be given to Chapple (1978) for having first made this change of scale which unquestionably opened new perspectives in the mechanical analysis of nappes and overthrusts. We have already seen the geometric characteristics of these thrust belts which are the premises of Chapple's mechanical model (triangular shape or wedge shape related to the opposing dips of the basal and upper surface slope, the presence of a decollement level, reduction in the displacement towards the front, and an increase in the deformation towards the rear, cf p. 11).

Chapple is in no doubt that the decrease in the deformation towards the front and the increase in the deformation towards the rear are the result of a horizontal push applied at the rear of the belt. Therefore he presents a resolutely compressive model and decides to treat the rocks in the basal layer and in the rest of the belt as an ideally plastic material. Regarding this issue, he is one of the rare geologists in the field of rock mechanics to recognise that the choice of rheology in the mechanical models is relatively arbitrary [14]. Chapple's mathematical theory is the first formulation of the critical wedge theory. Unfortunately, this theory of the critical wedge is presented in a way which is a bit muddled up and contains some contradictions or ambiguities that some of those who support the gravitational spreading model have not missed picking up on (e.g. Rod 1980).

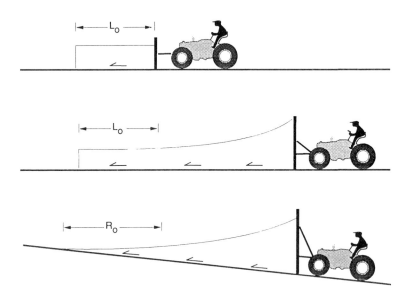

Fig. 32 - The solution of the mechanical paradox presented by Hubbert and Rubey (modified after Dahlen et al. 1984). Above: maximum length L_0 of a rectangular prism by rear compression. Beyond this length, the block cannot be translated within the framework of Hubbert and Rubey's theory. Middle: the theory of the critical wedge implies that a block of length greater than L_0 deforms at the rear until the surface slope sufficient to allow stable sliding is obtained. Below: the characteristic shape of the critical wedge in thrust belts. The front (length R_0) is a triangular zone without internal strain and with the upper surface slope close to the horizontal.

Yet, his principle is simple. The analogy with a bulldozer pushing snow is frequently used (Fig. 32). Through the action of the bulldozer, the snow is

---

[14] He writes, " The choice of perfect plasticity for the average rheology is a somewhat arbitrary one, but it does lead both to an easily understood characterisation of the material and to a solvable problem".

deformed, increasing its topographic slope, until a critical triangular or wedge-shape geometry is obtained. Therefore the wedge is displaced in a stable way, continually accreting snow at its front and simultaneously conserving its critical geometry which only relates to the conditions of stability of the displacement along the basal surface. Transposed to thrust belts, this model clearly stipulates that under the action of tectonic stresses applied at the rear, the belt will deform itself as a whole, through imbricated thrusts and/or internal strain. The slope of the upper surface slope will increase related to the deformation, until the geometry of the critical wedge is achieved. From this point the belt will stop deforming and instead will slide in a stable way along the basal surface. This theory is absolutely revolutionary because it demonstrates that the problem of the maximum length that a nappe can have when pushed at the rear is no longer an issue. For a given length (even semi-infinite!), there is always a corresponding critical wedge geometry, a simple relationship between the dip of the basal slope and the dip of the upper surface [15]. That is why this theory has equally been applied even at the scale of mountain chains (Platt 1986).

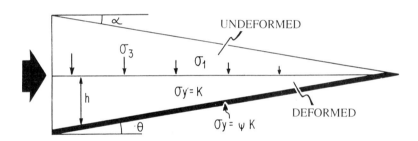

Figure 33 - Principal characteristics of Chapple's mechanical model (modified after Siddans 1984).

In Chapple's mechanical analysis, the wedge deforms in its lower part where the threshold of plasticity is of value K with a basal layer of low mechanical strength of which the threshold is ΨK ($\Psi < 1$). By considering the distribution of stresses in this lower part of the wedge, he calculates, at a point situated at a

---

[15] This theory is particularly elegant, " The once perplexing mechanical problems posed by the existence of large-scale thrust sheets and fold and thrust belts are, in a sense, an artefact of imposing a fixed rectangular geometry; this point has also been emphasised by Chapple (1978). Whereas Hubbert and Rubey calculated the maximum length overthrust attainable without a taper, we calculate the critical taper required for the emplacement of a thrust sheet of any length" (Davis et al 1983).

distance h from the base, the slope of the upper surface (α) needed in order to achieve the normal stress capable of deforming the lower part (Fig. 33). When the basal slope (θ) is small, the condition in order that deformation occurs in the basal layer, and so therefore the condition of allowing displacement of the wedge, is:

$$\rho g h \alpha + 2K\theta = \Psi K \qquad (37)$$

In comparison to the key equation of the model of gravitational spreading (equation 36), there is an additional term ($2K\theta$) which represents compression, inherent to the model. If the slope of the upper surface is insufficient, this term allows the lower part of the wedge to be deformed, therefore increasing the slope of the upper surface, until the conditions of sliding along the base are reached.

In a series of articles devoted to the theory of the critical wedge (Davis et al 1983, Dahlen et al 1984, Dahlen 1984, Lehner 1986), Chapple's initial work is taken up again by abandoning the hypothesis of an ideally plastic rheology (cf also Stockmal 1983 for the case of plastic behaviour) and in considering that the rupture is governed by the Navier-Coulomb equation (brittle deformation of Mohr-Coulomb type behaviour). The fluid pressure at the base and in the rest of the material is also taken into account but the cohesion of the rock, as in Hubbert and Rubey's work, is neglected. In this case, the relationship between the basal slope (β) and the surface slope (α) is linear and of the form:

$$\alpha + \beta = \frac{(1-\lambda_b)\mu_b + (1-\rho_w/\rho)\beta}{(1-\rho_w/\rho)+(1-\lambda)K} \qquad (38)$$

where λ_b and λ correspond to the fluid pressure at the base and in the rest of the belt, ρ_w and ρ the density of water and of the rocks of the belt respectively, μ the coefficient of friction at the base of the thrust belt and K a dimensionless number (cf equation (17) of Davis et al (1983) p.1159 for the exact formulation of this dimensionless number). As an example, for dry sand, this relationship becomes:

$$\alpha = 5.9° - 0.66\beta \qquad (39)$$

During laboratory experiments using dry sand and where different basal slopes are tested (Davis et al 1983) the material deforms until the critical wedge is obtained. At this point the experimental measurements which link the basal and upper surface slopes under stable conditions give a result which is very close to the theoretical prediction (Fig. 34):

$$\alpha = (5.7° \pm 0.20°) - (0.66 \pm 0.14)\beta \qquad (40)$$

However, this theory of the critical wedge, despite its degree of sophistication, is (as are all the others) highly dependent on the initial mechanical hypotheses. In particular, the taking into account of the cohesion of the material or again the removal of friction at the base of the wedge (i.e. decollement level with a pseudo-viscous layer) sensitively alters the conditions of stable sliding predicted by the model. Thus, experiments produced with sand resting on mercury (and therefore a total absence of friction) produce very narrow wedges and a large spacing of the structures (box folds and conjugate thrusts). The geometry of all of the structures together is then very close to the one observed in nature for the Jura or the Appalachians (Mulugeta 1988). Equally other experiments have demonstrated that the upper surface slope is clearly smaller than the one predicted by the theoretical model where the friction μ at the base of the models is less than 0.47 (Liu Huiqi 1992), the average value for the majority of sedimentary rocks. Likewise, some experiments with materials whose mechanical properties become modified during compression (for example an increase in the cohesion and the coefficient of internal friction) produce wedges which differ by several degrees to those calculated theoretically or which produce even, in certain cases, convex upper surfaces. Equally one notes that the wedge is not stable during its sliding and that it varies from 3 to 5° during the accretion of material at its front, suggesting that it slides whilst simultaneously changing its shape through internal strain which allows new material to be incorporated at the front (Mulugeta 1988).

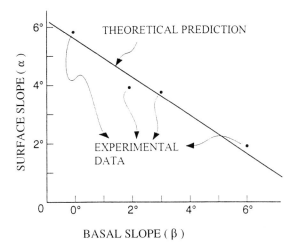

Figure 34 - *The theory of the critical wedge links the surface slope (α) and the basal slope (β) by a linear relationship (straight line). Results obtained from experiments have a good relationship with theoretical predictions (modified after Davis et al. 1983).*

It is equally correct that in this theory the role of gravity, the component due to the surface slope, is generally underestimated. Yet, it should be noted that the shear stress at the base of the wedge is equal to the total (equation (37) in the case of a plastic rheology) or to the product (equation (22) of Mulugeta 1988, in the case of a Mohr-Coulomb type behaviour) of two terms, of which one (ρgh) is entirely due to gravitational forces (cf equation 36). The importance of this term, considered to be sufficient alone in the model of gravitational spreading, is not really appreciated in the critical wedge theory, except as a necessary contributing force but never sufficient.

B. Fluid push model. The critical wedge theory has been elaborated with the intention of providing a mechanical explanation of the deformation and of the displacement observed in thrust belts in the external zones of mountain chains. However, in the case of the American Rockies, which is the natural field of application of this theory, the problem of the cause of the horizontal stresses applied at the rear of the belt has always been, and still remains, a controversial subject [16]. The subduction zone situated a hundred kilometres to the rear of the belt may not provide the necessary stresses for its activation, all the more than between the two a volcanic arc is situated occurring because of the rising of magmas created by the processes of the subduction zone itself. This intermediate zone is a zone of weakness, globally it behaves like a viscous liquid and because of this fact it prevents the transmission of stresses from the subduction zone to the thrust belt.

In 1981, the English geologist Smith proposed a solution to this dilemma by searching for the source of the compressions recorded at the rear of the thrust belt in the magma belt itself. It is the volcanic arc through the vertical intrusion of the magmas in the internal zone of the chain which creates the horizontal forces necessary for the formation of the thrusts in the adjacent external zone (Fig. 35). Rheologically, these magmas have a viscous behaviour, and therefore it is a hydrostatic pressure which is transmitted through the front of the diapiric plutons which pushes the sedimentary rocks towards the foreland of the chain. In otherwords, the *fluid push model* provides a relatively unexpected tectonic drive, for Chapple's mechanical model where the cause of the forces applied at the rear of the belt was fairly obscure. Thus, the geometry of the thrust belt is the same: a) wedge shape with a surface slope which produces a driving force and b) a basal layer of low mechanical strength. However, and it is a noteworthy difference with Chapple's model as with Elliott's model, the degree of internal strain in the belt is reduced to the maximum, the imbricated thrusts being assumed to propagate progressively towards the foreland, by never being active simultaneously. This means that the penetrative strain at the scale of the belt is only an illusion given by the finite state of the strain. This is because through time the deformation is active

---

[16] In criticising Chapple's model, Emile Rod (1980) comments ironically, "Here the question might be asked; how were the initial horizontal stresses produced? Chapple should have elaborated on what he meant by 'source of push' which he mentioned on page 1197. It seems that hidden somewhere in Chapple's model is Superman, who is pushing".

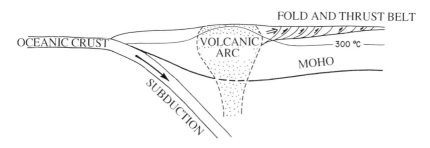

Figure 35 - The tectonic hypothesis of Smith's model of viscous compression: the rheological behaviour of the volcanic arc would be that of a viscous fluid creating a hydrostatic pressure at the rear of the thrust belt.

only along a single very narrow thrust surface, the displacement affecting the quasi rigid blocks (cf discussion p. 65). Finally, as in Elliot's mechanical analysis (1976), there is no hypothesis on the rheology of the rocks of the sedimentary prism except only that they behave as solids. The mechanical analysis is thus made from a simple analysis of the balance of forces acting at the base of the sedimentary wedge.

Looking at a natural case of the thrust of the Keystone-Muddy Mountain in Nevada (U.S.A.), Price and Johnson (1982) test Smith's model by making a mechanical interpretation of this overthrust considered to be the consequence of a back push caused by the intrusion of the enormous batholith of the Sierra Nevada. For this unit thrusting over two hundred kilometres, the mechanical analysis is made where the frontal wedge, the emergent ramp and the principal wedge (cf Fig. 38) are studied separately. The analysis demonstrates that only the front and the emergent ramp can follow a Mohr-Coulomb type behaviour and that the model is viable only if the principal wedge is considered as an elastic block overriding a basal layer of viscous behaviour. In Utah (U.S.A.), the Ruby's Inn thrust, of more moderate scale, has also been interpreted to be related to the intrusion of significant magmatic bodies (Merle et al 1993). The principal kinematics argument is the arc of a circle, over several tens of kilometres, which defines the pattern of the overthrusting on a map, a circular arc which surrounds a magmatic caldera of the same age as the overthrusting. Other examples of this type have equally been proposed for the displacement of sedimentary sequences during the emplacement of granites (Bouillin 1975).

The model of viscous compression has equally been tested through laboratory experiments whose convincing results seem to demonstrate the mechanical validity of this model (Merle and Vendeville 1995). In these experiments (vertical intrusion of a dense silicone into an overlying sequence of sand overriding a fine layer of less dense silicone), the thrusts are generated by two components which act simultaneously. The first component (Smith's model) is of viscous compression related to the ascension of the silicone simulating the magmatic intrusion (Fig. 36). The second (the model of Goguel, Kelhe, Hsu) is of gravitational gliding of the sand on the basal layer of lower mechanical strength (silicone) along the slope of the topographic dome produced by the vertical rising of the silicone which simulates the magma. In nature, this second component can be sufficient to cause some sliding of cover above an incompetent layer (cf p. 60 the case of Bearpaw Mountains).

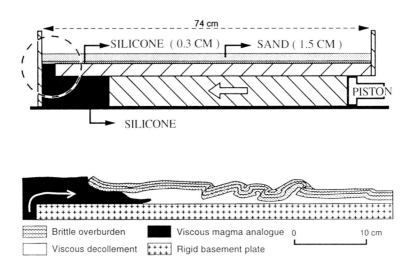

Figure 36 - Analogue experiment of thrusts induced by magmatic intrusions (Merle and Vendeville 1995). Above: Experimental set-up. The silicone (black) of the reservoir is compressed laterally and extrudes vertically into the sand layers overriding a fine layer of silicone, below: enlargement of the process of intrusion and compressive structures in front of the intrusion.

3.3   Composite Models

Several mechanical models analyse a prism of rectangular shape pushed from behind or resting on an inclined plane. By analysing again in the context of Hubbert and Rubey's paradox, a prism of triangular shape, instead of this rectangular block, can equally be represented. The maximum length that this block which is pushed at the rear can have is therefore double the length of a prism of rectangular dimension. This is because the mean vertical stress related to the weight of the nappe is two times lower (cf Price and Cosgrove 1991). It is an important point which must be kept in mind when the mechanical models are discussed which integrate a wedge-shaped geometry in their hypotheses, even if the theory of the critical wedge goes well beyond this simple fact.

It is equally odd that several authors close themselves within some models which are too restricted. For example, it is difficult to understand why Elliott in his paper of 1976 refuses to take into consideration the possibility of an exterior force applied at the back of the nappe or of the thrust belt [17]. On the other hand, Chapple considers gravitational forces to be always negligible, although in his model he absolutely needs a surface slope and therefore also a gravity component. As Elliott remarked in his response to Chapple's model (Elliott 1980, p. 186), and he recognised there implicitly that a composite model is envisigeable, gravitational forces increase linearly with the length of the overthrust, so that it is not absurd, in some instances, to envisage a change in the relative importance of the two compressive and gravity components.

It is Goguel (1948) once again, well before the others, who first understood that the gravity and compression models were not contradictory and could be combined without any difficulty. Analysing the displacement of a block of rectangular dimension with a basal layer of low mechanical strength and an ideally plastic rheological behaviour, he proposed three simple models: 1. sliding on an inclined plane, 2. a back push along a horizontal surface and 3. a back push on an inclined plane. In this third model, he thus simultaneously combined compression at the rear and gravitational gliding. Likewise, it is possible, by retaining a plastic rheology and a mechanically weak basal layer, to propose three other similar models by no longer considering a rectangular shaped prism, but a wedge shaped form (Siddans 1984). In all, still retaining the same hypotheses on the rheology of the material, eight variations can be proposed where the discriminating parameters are the two basal and upper surface slopes (Siddans 1984)[18].

---

[17] Or, at least, consider this force to be more or less without importance: "It would appear that most thrusts move essentially under the influence of stresses induced by the down-surface slope component. Compressive stresses set up by the hinterland pushing horizontally against its foreland are negligible in so far as entire thrust sheets are concerned" (Elliott 1976, p. 956).

[18] After Elliott (1976, p. 949) who talks of the "important but frequently overlooked treatment by Goguel" and Voight (1976, p. 157) who considers his work like a "milestone in the history of tectonic theory", Siddans was the third modern geologist to pay homage to Jean Goguel. He does it with humour: "If there were two prizes available for: the best mechanical model of Alpine thrust tectonics, the most ignored mechanical model of Alpine thrust tectonics, surely they should both be awarded to J. Goguel" (Siddans 1984, P. 281). More recently, Bayly (1992) has emphasised the delay that the recognition of Goguel's work has caused in the research on the mechanics of thrusting.

In what concerns the rheology of the material at the time of the overthrust, the same flexibility in reasoning must be seasonable. The fact is that the geometrical and rheological variations proposed by the researchers of rock mechanics can always be found in nature, thus it is true and indisputable that each nappe is a particular case with its own specificity. The geological environment is determining here and this environment varies equally through time, such as where the deformation can be ductile (pseudo-viscous behaviour) at the base of the nappe when deformation begins and can evolve into brittle deformation (Mohr-Coulomb type behaviour) towards the end of the displacement.

3.4 The Role of Erosion

We cannot know how to finish this review of the different mechanical models without calling to mind a parameter which, although having been fairly little studied, is of primary importance in the emplacement of nappes and of overthrusts. Let's examine the case of the McConnel overthrust studied by Elliott in 1976. The reconstructed displacement is of the order of 40 kilometres, the thickness of the thrust sheet is about four kilometres and the displacement rate is close to 5mm/year. At the front of the overthrust, the emergent ramp causes the sedimentary succession to displace from its original position (four kilometres in depth) up to the surface, which means to carry it on the back of the autochthon which in this case is an erosional surface. In other words, there would therefore be a mountain of four kilometres thickness (i.e. the thrust sheet) being displaced over forty kilometres along an erosional surface! Such an eventuality is obviously totally improbable. Naturally, during the eight million years during which the displacement of the McConnel thrust occurs, the front is continually eroded and the erosional detritus will be deposited in the sedimentary basin situated in front of the thrust, and even very far ahead of the thrust (Johnson 1981). Because of the erosion, it is reasonable that the shape of the front be wedge-shape, which as has been seen, reduces the normal stress needed to push along an erosional surface by a half (cf p. 72) (Fig. 37). In an extreme case where the rate of erosion is infinitely faster than the rate of displacement of the thrust, the front would be totally eroded as the thrusting occurs and paradoxically there would never be any real overthrusting on the autochthonous unit in front of the emergent ramp.

On the other hand, it should be remembered that all the models presented until now in terms of the simple rectangular shaped blocks resting on a planar surface, hardly represent an acceptable simplification of the reality. The rectangular prism which represents only the principal part of the moving unit is inevitably bounded at its front by an emergent ramp dipping at around thirty degrees. The thrust must again remount this slope to reach the surface. The resistance to the displacement is thus considerably greater than in the previous mechanical analyses. As this additional resistance is directly a function of the normal stress (related to the weight of the frontal part of the thrust) on the emergent ramp, the essential factor capable of reducing this resistance (which needs to be studied in detail) is the erosion itself. The geological reality is therefore found between two end members, the one where the erosion is so great that the front is totally planed off until the level of the neighbouring autochthonous surface, and the case where erosion is non-existent.

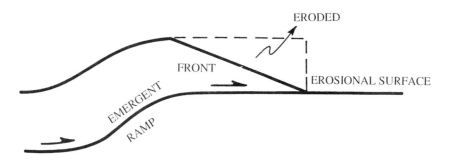

Figure 37 - The role of erosion on the emergent ramp: the front of the thrust is reduced by erosion to a wedge shape facilitating movement on the ramp (modified after Johnson 1981).

In 1963, Raleigh and Griggs analysed in detail the possible effect of the front in Hubbert and Rubey's theory of pore fluid pressure. The same mechanical hypotheses are used, namely a Mohr-Coulomb type behaviour, such as with the omission of cohesion (τ_0) and an abnormally elevated pore fluid pressure particularly elevated at the base of the thrust (λ_1) than at the level of the front (λ_2). In the hypothesis of gravitational gliding (Fig. 38), and in the case where the front is entirely eroded until the level of the neighbouring autochthonous surface, the resistance to the movement at the level of the principal block (of volume V_1, length x and thickness z) will be equal to the friction met along the base inclined at an angle of (θ) plus the force (F) necessary to push the front along the emergent ramp. The equation of equilibrium of the forces for sliding on a slope is therefore;

$$\rho g V_1 \sin \theta = (1-\lambda_1) \mu (\rho g V_1 \cos \theta + F \sin \theta) + F \cos \theta \qquad (41)$$

Likewise, the equilibrium equation of the forces for displacement on an emergent ramp is equal to;

$$F \cos \beta = (1-\lambda_2) \mu (\rho g V_2 \cos \beta + F \sin \beta) + \rho g V_2 \sin \beta \qquad (42)$$

where β is the slope of the emergent ramp, V_2 is the volume of the front and μ is the coefficient of internal friction ($\mu = \tan \Phi = 0.58$).

With a certain number of reasonable approximations (cf Raleigh and Griggs 1963, p.822) and by supposing that $\beta = 30°$ (the value which is most often observed for the slope of emergent ramps) equations (41) and (42) combine together to give:

$$\tan\theta = (1 - \lambda 1)\mu + (\frac{3z}{2x})(\frac{2 - \lambda 2}{2 + \lambda 2}) \qquad (43)$$

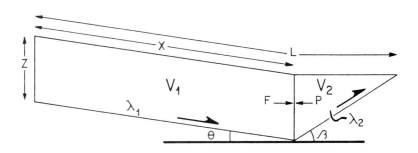

Figure 38 - Mechanical analysis of the resistance to displacement related to an emergent ramp in the case of gravitational gliding (modified after Raleigh and Griggs 1963).

This equation differs to the one of Hubbert and Rubey (equation 22) in that the term on the right is zero in the absence of an emergent ramp. If it is supposed that $\lambda_1 = \lambda_2 = 0.95$, the minimum slope possible to allow two blocks of 50 km and 100 km length to slide is 6.3° and 3.7° respectively. At a first glance, these values for the slope angles seem to be reasonable. They become much less so if a slope of 3.7° over 100 kilometres is considered. This would imply an altitude at the rear of the nappe close to 6500 metres using the hypothesis where the frontal ramp emerges to the surface at sea level. Could it also be possible that the fluid pressure can remain so elevated over hundreds of kilometres along the basal surface? If the erosion is considered to be inactive then the results are even worse. The volume of rock (V_2) implies that the front is therefore double and the necessary slope for gravitational sliding is equally doubled. Subsequently Raleigh and Griggs study the case of a rear push of a prism of rectangular dimension with a high degree of erosion at the front. In effectuating a mechanical analysis similar to the preceding one and by taking $\lambda_1 = \lambda_2 = 0.90$, the maximum length (L) of the rectangular prism is a function of the thickness (z) and is given by the formula;

$$L = 53.6 + 2.3\,z \qquad (44)$$

which, compared to the equation for the maximum length in the absence of a front (L = 53.6 + 10.4z), indicates a significant reduction in the length supported by the material. By way of example, a block of six kilometres thickness has its maximum length reduced by half in relation to the value calculated by Hubbert and Rubey. Naturally, in the absence of erosion, the displacement becomes even more difficult and the maximum length is again reduced. These results indicate above all the quasi impossibility of obtaining overthrusting if erosion does not substantially reduce the resistance created by the passage of the emergent ramp. All the more it should not be forgotten that Raleigh and Griggs's calculations are made by neglecting the material cohesion, an hypothesis which, as has been seen previously, is highly debatable.

Equally it should be indicated that the analysis of the stresses necessary to go up slope along the emergent ramp made by Raleigh and Griggs is incomplete. In effect, it is necessary to add the stress needed to produce the flexure of the block from the base to the upper surface of the ramp to the component due to the weight of the block on the ramp and to the component related to the resistance to the sliding along the ramp. This last component is far from being negligible as Wiltschko demonstrated in 1981.

However, other authors have contested that, as Raleigh and Griggs have considered it, the emplacement of thick thrusts emerging to the surface without erosion are excluded. By recognising that erosion largely facilitates displacement, Price and Johnson (1982) consider that the displacement of large overthrusts is possible even in the absence of erosion. It is true that their model, produced from the study of the Keystone Muddy Mountain overthrust (Nevada, U.S.A.) is based on some rheological hypotheses which are totally different to the ones used in Checkow and Merle's model (cf p. 75). Once again, the conclusions from the mechanical studies on the emplacement of nappes and overthrusts are very dependent on the hypotheses on the rheology of the material at the time of the thrusting.

However it may be, experiments produced on scale models in the laboratory, with an allochthonous sequence made up of brittle material (sand) overlaying a layer of low mechanical strength (silicone) along a flat, clearly show that erosion, by reducing the volume of material involved on the ramp, allows the overthrust to remain active in a quasi-unlimited way (Merle and Abidi 1995). Without erosion, the displacement rate reduces over time until movement along the ramp totally stops (Merle and Abidi 1995, cf also the numerical models of Beaumont et al. 1992).

3.5 Classification of Emplacement Mechanisms

This general survey of the mechanical studies devoted to the emplacement of nappes does not claim to be exhaustive. Nevertheless, the bulk of the perspective of the researchers of rock mechanics has been presented, so it is useful to extract some points from the maze of these different models to produce an outline which can be used as a footbridge between the mechanics and the kinematics points of view which we will approach in the fifth chapter. The hypotheses on which the mechanical models are founded centre around the following three points:

1. The nature of the forces which cause the displacement (forces of a gravitational or tectonic origin).
2. The rheological behaviour of rocks at the time of the thrusting.
3. The geometric form of the overthrusted unit (prism of triangular or rectangular dimension).

If these three points are considered indifferently, a fairly large number of models can be proposed and those of which that we have already covered represent only a small proportion of them. To propose a classification of the different emplacement mechanisms on the contrary consists of placing these hypotheses within a hierarchy in order to define the key mechanisms where all the mechanical models will naturally find their place. The nature of the forces causing the displacement is obviously the major hypothesis which generates two distinct mechanisms, gravitational or non-gravitational. In the end, all the mechanical models are classified as one or the other category.

The rheological behaviour of rocks at the time of the thrusting is equally an important issue but it is not as important as the previously mentioned hypotheses on the nature of forces of emplacement mechanisms. Its taking into account without discernment in the classification is a scattering factor which we should be aware of. By way of example, it can be stated that Goguel's model (ideally plastic rheology) or Kelhe's model (viscous rheology or power law) are two similar models which are based on the flowing of a basal layer of low mechanical strength in a gravitational gliding. It would appear more relevant to equate these two models by distinguishing them from gliding without deformation of a rigid block by brittle failure at its base. Likewise, in the case of gravitational spreading, the supporters of the model have not proposed any particular rheological hypothesis because the capacity of rocks to flow under their own weight is supposed to occur at the scale of the whole nappe without prejudgement of the type of deformation at the scale of the outcrop (cf the issue of the scale of observation, p. 65). Finally, the theory of the critical wedge has been developed with an ideally plastic rheology, then with a Mohr-Coulomb type behaviour; nobody would consider separating these two approaches into two distinct mechanisms.

The geometric shape is a parameter whose importance must be appreciated with the nature of the forces causing the displacement. In the case of a rear compression, it is of little importance to know if the compression is applied to a prism of rectangular or triangular dimension. Regarding the theory of the critical wedge, it intervenes above all to demonstrate that large scale structure can be displaced by rear compression. On the other hand, the respective role played by the basal and upper surface slopes in the gravitational models is a determinant geometric characteristic. It defines a boundary between two gravity mechanisms, spreading and gliding, which share nothing between them in common, except the role of gravity.

In total, four major emplacement mechanisms emerge from the research of those investigating the mechanics of overthrusts (Fig. 39). Without anticipating the chapter devoted to the kinematics, it should be emphasised that the identification of these four mechanisms is justified by their specific strain pattern which allows them to be recognised in the field. Naturally, these four mechanisms are borderline cases which can possibly be combined to give body to hybrid mechanisms.

1.   Rigid gliding. This is gravitational gliding which occurs along a favourable basal slope. The displacement is discontinuous through rupture at the base of the rock material, without any internal strain occurring in the rock mass during displacement. This rupture at the base can occur by hydraulic fracturing.

2.   Ductile gliding. The geometric conditions are identical to rigid gliding but the displacement is continuous, allowed by the flowing of a basal layer of low mechanical strength parallel to the basal slope.

3.   Spreading. This mechanism relates to a flowing of the nappe under its own weight. In reverse to the two preceding mechanisms, it is the slope of the upper surface, and not a hypothetical basal slope, which determines the movement of the material tending towards gravitational equilibrium.

4.   Rear compression. This mechanism reflects the tectonic stresses which, applied at the rear of the rock material, displace it along a more or less horizontal surface. A basal layer of low mechanical strength is needed for large displacement to occur.

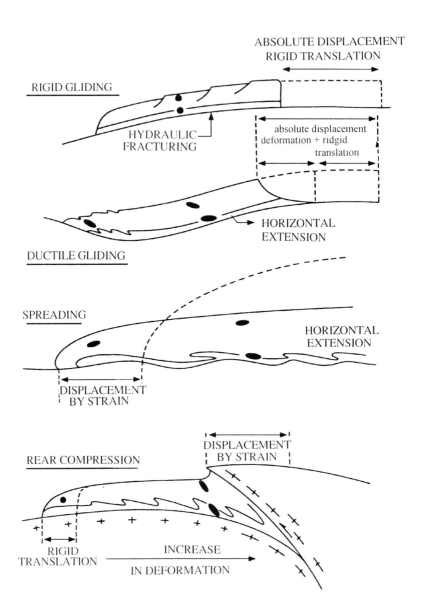

Fig. 39 - The four mechanical models for the emplacement of nappes; schematic representation in their natural environment.

CHAPTER 4

4 KINEMATICS

4.1 The Kinematics Approach

Kinematics is simply geometry in time (Goguel, 1948, p.7). To reconstruct the successive stages in the formation of a geological structure and the strain associated with each of these stages constitutes the foundations of the kinematics approach. For a long time, this approach remained content with intuitive and approximate reconstructions where the development of the structure was considered in its entirety without explicit reference to the type of internal strain which was associated with it. The transfer into the earth sciences of the concepts of the theory of strain during the 1960's (Ramsay 1967) has allowed, over a relatively short space of time, a considerable progress and a renewal in the field of investigation of structural geology. This transfer has opened out into the study of strain patterns which aims to determine the internal strain at the scale of the structure to emphasise its consistency and its logic according to the observed spatial variations. Thus, the internal strain can be synthesised at the scale of the whole of the mountain chain to determine its mode of formation (e.g. Choukroune and Seguret 1973), at the scale of the internal or external zones of a chain (e.g. Choukroune and Gapais 1983) or naturally at the scale of a simple geological structure such as a diapir (e.g. Brun

1983) or a nappe (e.g. Wood 1973, Milnes and Pfiffner 1980, Ramsay 1981, Siddans 1983). This approach aims to be predictive. Based on the hypothesis that the mode of formation of a structure produces a specific and clearly identifiable strain pattern, it attempts to determine, experimentally and numerically, the strain patterns of each of the possible modes in order to give the keys which will allow them to be recognised in their natural environment.

In what concerns nappes and overthrusts, the four models stemming from the mechanical studies approached in chapter 3 (Fig. 39) are end member cases whose principal characteristic is to present an invariant internal strain whatever the mechanical approach chosen to study them. Concretely, in two dimensions, the state before deformation is represented by a grid comprising square elements (representing the absence of preferential lengthening and shortening) and the deformed state corresponds to the distortion of each of the initial squares (Fig. 40) where the small axis gives the value and the direction of the maximum shortening. Thus a kinematics approach to the emplacement of nappes and overthrusts is defined, which consists of analysing the relationships between the displacement and internal strain through time. In the end, this approach attempts to reconstruct and to understand the strain produced during the displacement of allochthonous units. As a paradoxical example, the internal strain in the case of a rigid gliding is visualised by the square element grid itself since, by definition, the displacement takes place without any internal strain (Fig. 40).

The most important kinematics criteria which must be defined in the laboratory, and then researched in the field to determine the emplacement mechanism of an allochthonous units are the following:
1. The schistosity trajectories in cross-section (parallel to the displacement).
2. Strain gradients, both in the vertical and from the front towards the rear.
3. The strain regime: coaxial or non coaxial.
4. The relationships between the stretching direction and the displacement direction.

Experimental studies show that the boundary conditions determine, for a large part, the finite strain pattern associated with the displacement of allochthonous units. Thus, the study of the internal strain in gravity nappes slightly alters the mechanical synthesis into four major models. In effect, if ductile gliding is possible for rocks flowing under their own weight, and if the whole of the nappe (and not only a basal layer) suffers such a strain, it is less than reasonable that it will escape a vertical shortening, which means a gravitational spreading component. In

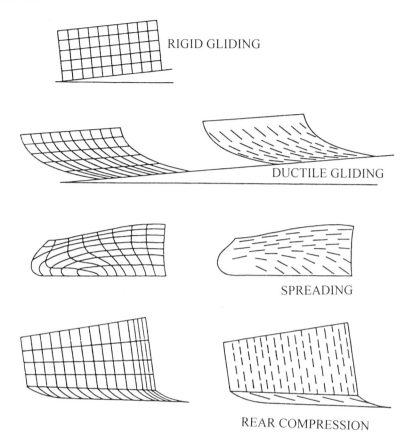

Figure 40 - Grids of initially square elements representing the internal strain in each of the four mechanical models defined in chapter 3. To the right: pattern of maximum stretching trajectory.

other words, a material flowing under its own weight on a favourable basal slope must combine gliding and spreading. This leads to the concept of gliding-spreading being introduced, whose content is fairly close to the one of flowing through gravity introduced by the Alpine geologists in the 1940's (Schneegans 1938, Gignoux 1948). On the other hand, the horizontal shortening of the basement rock found in the majority of collisional chains causes the extrusion of sedimentary material situated above. This extrusion is essentially lateral because as it is controlled by gravity, it does not allow significant vertical extrusion of the material. This combination of compression and gravity constitutes the extruding-spreading model. From the point of view of internal strain, the model of gravitational spreading must be subdivided into three models, which present sufficiently distinct strain patterns to be clearly identifiable in the field (Fig. 41):

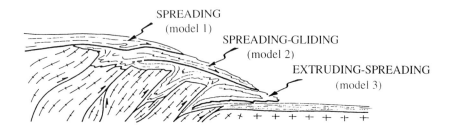

Figure 41 - Schematic representation of the three types of gravitational spreading in their natural environment.

1. **Spreading model** (sensu-stricto): the upper surface slope determines the movement of the material and the basal slope is of zero inclination, or slightly inclined in the direction opposite to the displacement.
2. **Gliding-spreading**: the upper surface slope and the basal slope are inclined in the same direction and spreading is combined with ductile gliding.
3. **Extruding-spreading**: sedimentary material is extruded following an intense shortening of the underlying basement and spreads in the direction of the topographic slope.

4.2 Theoretical and Experimental Strain Patterns

4.2.1 STRAIN FACTORISATION.

It should be pointed out immediately that these last three models just like the models of rear compression and of ductile gliding, are studied experimentally by considering total adherence of the allochthonous unit along the basal surface. This relates to some conditions approaching natural examples, where it is known that the shear component increases in a spectacular manner towards the base, and therefore towards the fault contact (e.g. Williams 1978, Milton and Williams 1981). In nature, a part of the displacement can naturally be assured through a component of rigid translation which does not alter the internal strain observed in experiments.

The other case, relating to a total absence of friction along the base (cf Ramberg's experiments 1981), is not considered to be realistic from the geological point of view and besides has never been observed in nature.

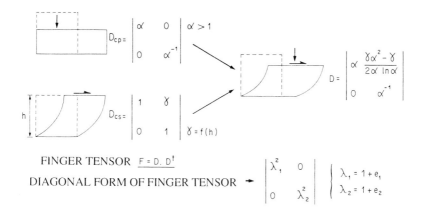

Figure 42 - Example of a two-dimensional model: factorisation of the internal strain into two components (α and γ) in gravitational spreading. The strain tensor D corresponds to the simultaneous combination of these two components. The Finger tensor ($D.D^t$) allows the different parameters of the strain ellipsoid to be calculated.

In two dimensions, the experimental study is coupled with a mathematical analysis which consists of dissociating the internal strain into two components (e.g. Ramberg 1975a and b, Sanderson 1982) (Fig. 42):
1. A component of *pure shear* where the stretching α is parallel to the basal plane and relates either to a vertical shortening and a horizontal lengthening ($\alpha>1$) or to a horizontal shortening and a vertical lengthening ($\alpha<1$).
2. A component of *heterogeneous simple shear* whose shearing plane is parallel to the basal plane and which is characterised by an increase in the intensity of shearing towards the base of the nappe ($\gamma = f(h)$, Fig. 42).

Schematically, in two dimensions, it is considered that the simple shear component γ is identical for all of the models, weak or absent at the top and increasing progressively towards the base where considerable values are reached. On the other hand, the pure shear component α is the one which allows the different models to be differentiated since it takes a value greater than 1 for a vertical thinning (i.e. gravitational spreading), a value equal to 1 for an invariant thickness (i.e. ductile gliding) and a value less than 1 for vertical thickening (i.e. rear compression). These are thus the relative proportions of γ and of α from the top to the bottom and from the rear towards the front of the allochthonous unit which allows the different strain patterns observed in the principal emplacement models to be explained (Sanderson 1982).

It is possible to calculate a graph where the strain intensity and the angle between the long axis λ_1 of the strain ellipsoid and the basal plane (which in two dimensions means the dip of the schistosity) are reported according to the curves α and γ constants (Coward 1980, Sanderson 1982, Merle 1986, Tikoff and Fossen 1993) (Fig. 43). Using this graph, from the angle between the schistosity and the basal plane the exact proportions between the two components α and γ can be determined and which allows firstly, the mechanisms of emplacement to be eliminated which do not relate to the value α observed (greater, less than or equal to 1) for these models. Secondly, it is generally stated that, for a given model, the value of α is constant in the vertical of the allochthonous unit (Merle 1986). This important point allows the schistosity trajectories of the two dimensional models to be explained in a theoretical way, as we will see later.

These two components are simultaneously combined (Fig. 42). The state of strain at each point can therefore be obtained by calculating the eigenvalues and the eigenvectors of the Finger tensor [19]. This mathematical approach allows the theoretical strained grids to be calculated corresponding to the studied model. Thus the validity of the mathematical models used can be tested, by comparing these theoretical grids to the grids obtained experimentally.

The three dimensional experiments tackle more complex problems related above all to the relationships between the stretching of the material and its displacement. As an example, three dimensional gravitational spreading is characterised by a divergent or radial displacement of the material, similar to the one observed at the front of a glacier at the exit of a two dimensional channel (e.g. Ramberg 1964).

---

[19] The simultaneous combination D of the two components D_{CS} and D_{CP} is calculated from the strain rate tensors (Hsu 1967, Ramsay 1975). The Finger tensor is obtained by multiplying the strain matrix D by its transpose D^T (Malvern 1969, p. 174, De Paor 1983). The eigenvalues of this tensor are equal to the quadratic elongations (λ^2) and the eigenvectors give the orientations of the three axes of the ellipsoid.

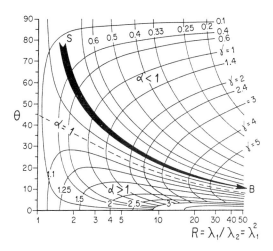

Figure 43 - Graph allowing the exact proportions of the two components α and γ to be quantified from the axial ratios of the strain ellipse (R.) and from the angle between the schistosity and the basal plane of the allochthonous unit (θ) (after Merle 1986, Tikoff and Fossen 1993). The concave trajectory of the schistosity in the model of rear compression is interpreted on the constant α curve (arrow) by the increase in the component γ from the top (S) towards the base (B) (explanation in the text).

Equally the mathematical analysis of a radial gravitational spreading occurs by a dissociation into two components (Fig. 44):
1. A component of *heterogeneous simple shear* marked by an increase of shear towards the base. The directions of shear are radial and therefore parallel to the direction of displacement, equally radial.
2. A component of *pure shear* characterised by a vertical shortening, identical to the one observed in two dimensional experiments, but also a radial horizontal shortening, parallel to the directions of displacement.

In the case of extruding-spreading, the horizontal compression causes a thickening at the rear of the nappe and therefore a surface slope which favours a spreading process. At the back, the strain components are therefore similar to the ones for radial spreading, except that a vertical extension is found and not only a vertical shortening (Fig. 44). Towards the front, the effects of the horizontal compression decreases, and the model of extruding-spreading is exactly similar to the model of radial spreading. Expressed in another way, the process of extrusion and spreading interact, and zones of spreading strictly said often become integrated within wider domains, whose rear is a zone of extrusion. Therefore these two models can be analysed together by making the vertical component of the internal strain vary, from the rear towards the front of the nappe.

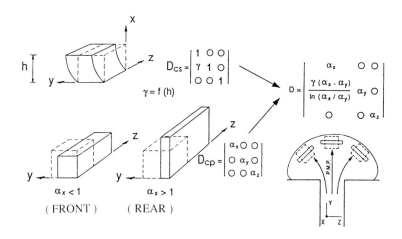

Figure 44 - Example of three dimensional model: factorisation of the internal strain into two components α and γ in the extruding-spreading. The pure shear component α_x exceeds values greater than 1 at the rear and values less than 1 towards the front. The tensor D relates to the simultaneous combination of the two components of strain. Bottom right: surface view of an extruding-spreading experiment showing the radial trajectories of the particles (PMP) and the concentric stretching perpendicular to the displacement (transformation of the initial squares (dashed) into rectangles). System of square co-ordinates to analyse the internal strain.

This dissociation of the internal strain into two components is defined under the term of *strain factorisation*. Since its introduction into geology by means of theoretical analyses (e.g. Ramberg 1975, Sanderson 1982), this method is equally frequently used on natural examples and helps in the interpretation of data of internal strain collected in the field (e.g. Sanderson et al. 1980, Coward and Kim 1981, Kliegfield et al. 1981, Beach 1982, Law et al. 1984, Evans and Dunne 1991, Seno 1992).

Kinematics

4.2.2 REAR COMPRESSION.

In two dimensions, this model relates to the case where the pure shear component α is less than 1. Therefore it relates to a thickening of the allochthonous unit towards the rear, which means towards the source of the compression. Numerical simulations (Guillier 1988), experimental modelling (e.g. Vialon et al. 1984 Guillier 1988) and field studies (Siddans 1979) give very comparable results. The numerical simulations introduce a gradient in the component α which decreases progressively from the rear towards the front. The gradient of the shear component relates to an increase of γ from the top, where its value is zero, until the base where its value is maximum.

The major characteristics of internal strain which are found in the field are the following (Fig. 45):

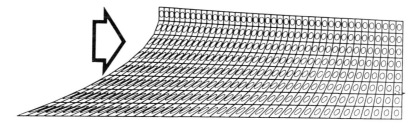

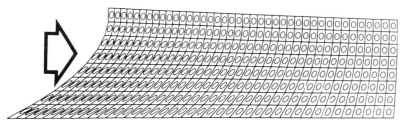

Figure 45 - Deformed grids obtained numerically from the distribution of the two components α and γ in the two dimensional models of rear compression (after Guillier 1988). Note the two strain gradients relating to an increase in strain from the top towards the base and from the front towards the rear.

1. The principal strain gradient relates to an increase in intensity of strain from the front towards the rear. The front of the model is unaffected by internal strain and is translated rigidly. A clear increase in the deformation from the top towards the base is equally observed, following the example of all the nappes whatever its emplacement mechanism.

2. The strain in the upper part of the nappe is close to a coaxial regime (negligible γ), characterised by a horizontal shortening and vertical lengthening. At the base, the strain is similar to very intense simple shear (α is negligible in relation to the intensity of γ).

3. The schistosity trajectories in a section parallel to the displacement are of concave type: sub-vertical in the major part of the nappe and gently dipping towards the base.

It should be noted that the component α produces vertical schistosities whilst the γ component produces schistosities which, dipping at forty five degrees to the first increment of strain, become horizontal as a function of the intensity of the strain. The dip of the schistosity at each point is directly a function of the spatial distribution and of the value of these two components in the allochthonous unit. Analogue models produced in the laboratory show that the values α are, on a given vertical, practically constant. For this reason, the trajectory of the schistosity in section can be explained by positioning itself on a theoretical constant α curve which reveals, through the increase of the value γ of simple shear, from the top towards the base, a progressive reduction in the dip of the schistosity (Fig. 43).

Outside the classic geometry of ramps and flats, other structures formed include folds whose axial plane follows the concave form of the principal plane of flattening of the material (i.e. the schistosity). The overturning of folds is therefore more marked at the base of the allochthonous unit while the summit comprises upright folds associated with a vertical schistosity. This similarity between the axial planes of folds and the dip of the schistosity is well visible in experiments on scaled models produced in the laboratory (e.g. Dixon and Tirrul 1991) and are found in numerous allochthonous units (e.g. Sanderson 1979, Mitra and Elliott 1980, Rattey and Sanderson 1982, Loosveld and Schreurs 1987, Gray and Willman 1991).

It is important to note that this fold geometry is different in fold and thrust belts such as the Jura (Laubscher 1975) where a decollement zone of low mechanical strength exists at the base. In this case, the folds are less often overturned and show the characteristic box fold form. This difference is explained in terms of the separation of the components of the internal strain, or *strain partitioning*. The component γ is concentrated in the basal level producing a strain close to simple shear and is practically absent in the upper competent part which records only the effects of the α component. This upper part, by far the thicker part therefore deforms through sub-vertical axially planar folds.

On geological maps, an arcuate shape of the allochthonous unit or the whole fold and thrust belt is often observed, which reveals that maximum displacement occurs in the central region and dies out along strike towards both lateral sides. We shall return to this point when studying lateral boundary effects.

4.2.3 DUCTILE GLIDING.

Although the model of ductile gliding can be applied to an entire nappe, it is likely that above all it concerns the strain in the levels of low mechanical strength at the base of certain allochthonous units (Kelhe 1970). The upper part of the nappe situated above this basal level is therefore translated rigidly and only the internal strain at the base of the unit is related to ductile gliding strictly said. The initial mechanical model (cf chapter on the mechanical models) assumes a flowing of rocks parallel to the basal surface, without thickening or thinning. By definition, the internal strain relates to heterogeneous simple shear with an increase of γ towards the thrust fault. In terms of strain factorisation, the α component of pure shear is equal to 1.

From the characteristics of simple shear the following kinematical criteria can been deduced:
1. The schistosity is of a concave form, set upright towards the summit and layered at the base. However, in reverse to the preceding model, the dip of the schistosity never exceeds forty five degrees since it follows the curve α =1 of the theoretical chart (Fig. 43).
2. As the model is two dimensional (by definition), the strain ellipsoid measured in the field must be of plane strain type (K = 1)[20] and the direction of the stretching lineation point to the direction of the displacement (again it must be noted that a component of rigid translation can occur in a completely different direction).

---

[20] K is the Flinn parameter (1962) which defines the shape of the strain ellipsoid. K = $[(\lambda_1 / \lambda_2) -1]/[(\lambda_2 / \lambda_3)-1]$ with $\lambda_i =1+e_i$ (value of the three axes of the strain ellipsoid) so that the ellipsoid is of flattening type when K is less than 1 (λ_1 and λ_2 greater than 1), of plane strain type when K is equal to 1 ($\lambda_2 =1$) and of constriction type when K is greater than 1 (λ_2 and λ_3 less than 1).

4.2.4 GLIDING-SPREADING.

This model combines ductile gliding and gravitational spreading. It therefore relates to the combination of a component of horizontal lengthening ($\alpha>1$) and a component of heterogeneous simple shear ($\gamma = f(h)$). In two dimensional experiments produced in the laboratory (Brun and Merle 1985, Merle 1986), displacement occurs due to a frontal rolling where the (fast) upper part of the front advances by overriding the (slow) lower part, a process which has already been mentioned in the chapter devoted to the geometry and to which we will return when studying the spreading model.

As in the experiments of the flow of viscous fluids on a basal slope (Huppert 1982), the upper surface (of dip equal to the basal surface at the beginning of the experiments) tends towards the horizontal, which causes a thinning at the rear and a thickening at the front. Two strain gradients are therefore observed: an increase of the strain from the rear towards the front and from the top towards the base. The strain progressively decreases towards the back so that a model can be imagined where the gliding-spreading nappes passes continuously to autochthonous material at its back (Fig. 46). This horizontal gradient is an important criteria allowing this model to be differentiated from the one of horizontal compression. It is often, and with just cause, the principal argument used in field studies to choose between rear compression or gravitational gliding (e.g. Siddans 1979, Durney 1982).

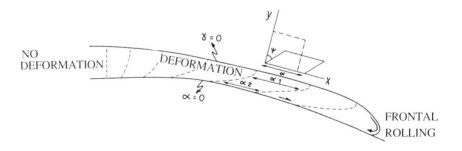

Figure 46 - Principal characteristics of the model of gliding-spreading. The displacement is ensured by a frontal rolling and the dashed lines represent the deformation of initially vertical lines. The displacement gradient is such that the allochthonous unit can link in continuity at the rear to the autochthonous domain. The internal strain is schematically represented by a combination of pure shear ($\alpha >1$) and simple shear (γ).

Strain analysis of experimental grids (again) shows that the component α is constant in the vertical and that the schistosity trajectories (in section parallel to the displacement) are explained owing to variations in the component γ from top to bottom and from the rear to the front of the allochthonous unit (Merle 1986)[21]. A double evolution, spatial and temporal, is seen. At the beginning of the experiments, three schistosity trajectories are observed: convex trajectories at the rear, sigmoidal trajectories at the centre and concave ones at the front (Fig. 47). As the experiment continues, the convex trajectories transform into sigmoidal trajectories and the sigmoidal trajectories transform into concave trajectories. At an ultimate stage, all the trajectories are concave. This spatial distribution and this temporal evolution of the trajectories is explained by the evolution of the two components α and γ from the top to the bottom and from the rear towards the front during time. Again, as for the models of rear compression and ductile gliding, the shape of the curves of the α constants allows both the three observed trajectories and also the progressive passage from one to the other to be taken into account according to the increase of the component γ from the summit towards the base (which explains each of the three trajectories) and from the rear towards the front (which explains the transformation from one to the other) (Fig. 47).

The shape of the trajectories in a vertical section is the best criteria to demonstrate an emplacement mechanism by gliding-spreading. Some sigmoidal trajectories have been observed in natural examples such as the Graz nappe in Austria (Ratschbacher et al. 1991) and low angle concave trajectories are frequently described in certain glaciers (Hudleston and Hooke 1980). This observation, coupled with the increase of the strain towards the front, is enough to eliminate spreading or rear compression and to opt for the gliding-spreading model.

4.2.5 SPREADING.

The displacement in the spreading model (as in the model of gliding-spreading) is produced by a frontal rolling. Gravitational spreading (in a wide sense) is therefore always made up of a fold-nappe at the front and a nappe at the rear. It is advisable

---

[21] In all strictness, the component α may not really be constant in the vertical because of the increase of γ from the rear to the front which necessarily brings about a variation of α (Fig. 46) (cf Merle 1986). Therefore it is more precise to say that the vertical variation of α is negligible in relation to the vertical variation in γ.

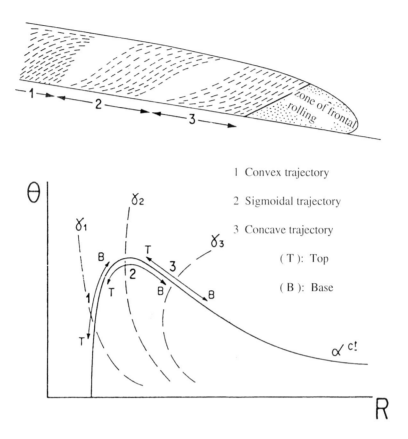

Figure 47 - The trajectories of schistosity in section parallel to the displacement in the gliding-spreading model. Above: the three trajectories (convex-sigmoidal-concave) observed from the rear towards the front. Below: interpretation of the trajectories from the α constant curves of the chart from Figure 43 (explanation in the text).

therefore, and this releases us from the purely geometrical classification described in the first chapter of this book, to put forward the presence of a frontal rolling (or fold-nappe) as a criteria to recognise spreading (or gliding-spreading). The internal strain is the same in the recumbent fold (with the exception of the frontal hinge) than in the principal body of the nappe. Even if the fold-nappes sometimes have a relatively weak internal strain (e.g. Potts, 1982), this frontal fold has some well defined characteristics which allow it to be recognised in the field and which allow it to be differentiated from recumbent folds produced by buckling (Fig. 17) (Merle 1990):

1. The sense of shearing is the same in the reverse and normal limb. This sense of shear conforms to the sense of displacement of the nappe.
2. The schistosity is parallel to the stratification in the frontal hinge as a result of the rolling process.
3. The schistosity trajectories (in section perpendicular to the fold axis) never reveal a fan-like pattern on either sides of the axial plane.

Gravitational spreading has been studied using two and three-dimensional models. The principal results of these two types of experiments are not comparable, which means that the boundary conditions influence in a determining way the internal strain of the models, as we have just had the opportunity of indicate. Knowing the emplacement mechanism is therefore not sufficient, in itself, to suitably predict the structural characteristics of the spreading models. There again, it should be considered that the two types of experiments are end members, so that natural examples are susceptible to present specific characteristics intimately related to their boundary conditions. Inversely, a good knowledge of the internal strain of a nappe must allow some reasonable hypotheses regarding the boundary conditions at the time of the emplacement.

Following experiments produced in the laboratory (Merle 1989), the following points will be principally retained as field criteria for gravitational spreading in two dimensions (Fig. 48):
1. The intensity of strain increases from the rear towards the front of the nappe, but above all from the summit towards the base.
2. The distribution of strain regimes is not anyhow. The strain relates to pure shear in the upper part of the nappe and to simple shear in the lower part.
3. The schistosity trajectory (in section parallel to the displacement) which results from the distribution of the strain regimes is convex: flat at the summit and inclined towards the base (without ever exceeding forty five degrees). In the ultimate stages of the strain, this convex trajectory is less and less perceptible, because of the intensity of the simple shear, and tends toward the horizontal. In reverse to the gliding-spreading model, concave or sigmoidal trajectories are never observed.

96   *Emplacement Mechanisms of Nappes and Thrust Sheets*

Three dimensional experiments (e.g. Gilbert and Merle 1987, Merle 1989, Cobbold and Szatmari 1991) attempt to understand the relationship between the internal strain and the divergent or radial movement of the material. Such divergent movements are observed in nature, sometimes even at the scale of a thrust belt (e.g.

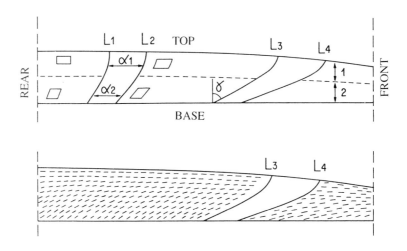

Figure 48 - The internal strain in two dimensional gravitational spreading. The deformation of the initially vertical lines (L1, L2, L3 and L4) shows the two strain gradients: increasing the intensity of strain from the rear towards the front and from the summit towards the base. Pure shear at the summit and simple shear at the base. Below: convex schistosity trajectories.

Crosby 1969). This problem is relatively complex, the experimental conditions are the most simple possible to demonstrate the key parameters which control the internal strain: at the exit of a channel where the silicone flows in conditions similar to two dimensional experiments, the material is left to spread radially on a surface equally plane and without borders. Through the observation of a grid of initially square elements which records the internal strain during the experiments at the surface of the model, it is stated that this grid records a strain through pure shear relating to a concentric stretching and radial shortening. In other words, and it is a primordial characteristic of spreading in three dimensions, the lengthening direction on the surface is perpendicular to the radial direction of the displacement (Fig. 49).

The component of pure shear relates to (i) a vertical shortening, (ii) a horizontal and radial shortening, parallel to the displacement vectors and (iii) a concentric stretching perpendicular to the displacement vectors (Fig. 44). This component of the internal strain is counterbalanced by a component of simple shear which becomes amplified towards the base of the model. This shear direction is parallel to the displacement vectors. In reverse to the pure shear component, this simple shear component produces a maximum lengthening of the material which is radial and no longer concentric.

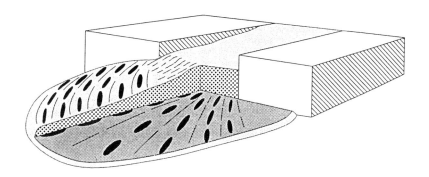

Figure 49 - Three dimensional spreading model: concentric stretching perpendicular to the displacement towards the summit, radial stretching parallel to the displacement towards the base (explanation in the text).

Thus, it is the relative part of the two components of pure shear and simple shear which determine, within the model, a concentric or radial stretching. Knowing the strain gradients in the spreading process therefore allows a spatial distribution of the concentric and radial stretching to be proposed. Stretching is concentric towards the summit of the nappe where the simple shear component is quasi non-existent. It becomes radial at the base where the component of simple shear reaches very significant values (Fig. 49).

Mathematical analysis integrating the variations of the two components of internal strain reveals two important points:

1. The passage between the two stretching directions is instantaneous, without transitionary direction, through simple permutation between the axes λ_1 and λ_2 of the strain ellipsoid. The obligatory stage between radial and concentric stretching is a uniaxial oblate ellipsoid (K = 0). On a Flinn diagram (Fig. 50), these variations can be followed: the strain ellipsoid is of constriction type at the summit of the nappe, then the increase in the component of simple shear makes the ellipsoid move into the field of flattening until the instantaneous stage of a uniaxial oblate ellipsoid (obligatory passage between concentric and radial stretching), then again in the field of flattening.

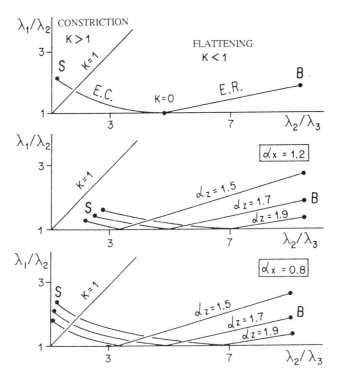

Figure 50 - Flinn diagram showing the changes of shape of the strain ellipsoid in three dimensional models of spreading and of extruding-spreading. Above: schematic representation of the general evolution observed from the summit (S) towards the base (B). Passage by a uniaxial oblate ellipsoid around the shortening axis (λ_3) between concentric stretching (E.C) and radial stretching (E.R) Centre: in the case of extruding-spreading ($\alpha_x > 1$), the ellipsoids stay in the field of flattening. Below: in the case of spreading ($\alpha_x < 1$), the ellipsoids are in the field of constriction at the summit of the nappe.

2. The schistosity is sub-vertical towards the summit and decreases rapidly towards the base where the component of simple shear layers it towards the horizontal (Fig. 51). The schistosity trajectory is concave, and no longer convex as in the case of two dimensional spreading. Sigmoidal trajectories are observed at the extreme front of spreading (Fig. 51). But what is the most important to note is the quasi-verticality of the schistosity towards the summit, a result which goes against the largely widespread idea that schistosity is always very flat in gravitational nappes(see Merle 1989).

4.2.6 EXTRUDING-SPREADING.

In three dimensions, this model presents some characteristics close to three dimensional spreading. Equally a horizontal shortening parallel to the radial displacement is observed and a concentric lengthening perpendicular to the displacement. On the other hand, the compression at the rear of the material causes a vertical thickening which emphasises the process of extrusion (Fig. 44). Thus the surface slope created facilitates the spreading process which, towards the front, takes over from the extruding process. That is why, as we have already previously indicated, the internal strain at the front of a zone of extrusion is basically similar to the one of spreading.

Taking into account this rear thickening in the numerical simulations allows the differences between extruding-spreading and spreading to be clarified (Fig. 51):
1. The stain gradient indicates, as in all nappes, an increase of strain from the summit towards the base, but equally from the front towards the rear, which means towards the primary cause of the thrusting.
2. The schistosity trajectories are concave and highly upright towards the summit of the nappe, very similar to the one which can be observed in models of rear compression. Towards the front, they pass progressively to the trajectories described for three dimensional spreading.
3. The strain ellipsoid of constriction type ($K>1$) is no longer observed. In all areas the ellipsoid is of flattening type ($K<1$), except at the passage between concentric stretching and radial stretching where the uniaxial oblate ellipsoid is observed ($K=0$).

It can be asked, with just cause, if experimental modelling and numerical simulations do not distance themselves too much from the geological reality that they are supposed to describe. In order to better define their interest, it is useful to

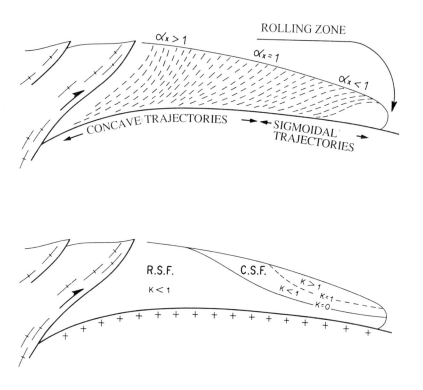

Figure 51 - Characteristics of the internal strain in the extruding-spreading model (whole of the figure) and of spreading (right side of the figure where $\alpha_x < 1$). Above: schistosity trajectories in section parallel to displacement. Below: Spatial distribution of the Flinn parameter K of shape of the strain ellipsoid (CSF: concentric stretching, RSF: radial stretching).

consider a famous natural case (the one of the Helvetic nappes) and to show how the method of strain patterns allows an appropriate model of emplacement to be selected. The Helvetic nappes relate to the stacking of several fold-nappes: the one located in the lower position (the Morcles fold nappe) is complete and bedded to the horizontal (Lugeon 1902, De Loys 1928, Badoux 1972) (Fig. 52). Several emplacement models, each very varied to the other, have been proposed (e.g.

Durney 1982, Ramsay et al. 1983, Butler 1985, Gillcrist et al. 1987, Burkhard 1988, Merle 1989, Dietrich and Casey 1989, Ratschbacher et al. 1991). The important points of the strain pattern in the Helvetic nappes are the following:
1. The intensity of strain increases considerably from the summit towards the base but equally from the front towards the rear (Ramsay 1981, Durney 1982).
2. The schistosity trajectories in a section parallel to the displacement are concave, very upright towards the summit and flat towards the base. Some sigmoidal trajectories are described towards the front of the edifice of the nappes (Ratschbacher et al. 1991).
3. The stretching lineation is parallel to the displacement in the lower part of the edifice of nappes and perpendicular in the upper part (Ramsay 1981).
4. These two domains are separated by a zone where the lineation is no longer detectable on the plane of schistosity, which means that the strain ellipsoid is of uniaxial oblate type (K=0) (Ramsay 1981, p. 302).

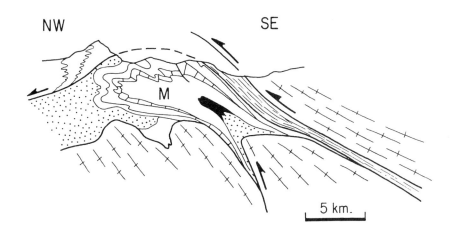

Figure 52 - The Morcles nappe (M) results form the extruding-spreading of sedimentary material during the shortening of the underlying basement (crosses) (modified after Jeanbourquin and Goy-Eggenberger 1991).

On the other hand, the upper part of the edifice of nappes is characterised by a constriction type ellipsoid (K>1) where the single stretching axis is perpendicular to the displacement. Everywhere else, except at the base of the Morcles nappe (K=1), the ellipsoid is of flattening type (K<1).

This strain pattern is sufficiently close to the one described for the model of extruding-spreading so it is hardly necessary to argue further to propose tectonic extrusion of the Helvetic nappes following the deformation of the underlying basement, and their synchronous gravitational spreading towards the more external zones (Durney 1982, Merle 1989, Ratschbacher et al. 1991).

A similar model can be proposed for some nappe edifices verging south of the Pyrenees (the Gavarnie-Mt Perdu nappes) where the internal strain characteristics (Seguret 1972, Choukroune and Seguret 1973, Deramond 1979, Deramond and Rambach 1979) can be interpreted as the combined result of compression and of gravity following a model close to extruding-spreading (Merle 1989). Yet it is necessary to demonstrate in the case of the Pyrenees (as for any such allochthonous unit) that all the internal strain is produced during the thrusting and is not associated in part to deformation episodes subsequent to the thrusting, as N. T. Grant (1992) suggests who proposes a relatively late stage of gliding-spreading for the nappes of the Pyrenees. Likewise, the stretching lineation in the fold-nappes in the Montagne Noire, France which are most often orientated perpendicular to the direction of displacement (Arthaud 1969 and 1970, Harris et al. 1983) could be related to a component of gravitational spreading during the emplacement (P. Choukroune, personal communication 1992).

4.3 Displacement Trajectory

To retrace the exact path of an allochthonous unit, from its original homeland up to its current position, is one of the most delicate problems in the kinematics analysis of nappes. The problem is only posed in the case of large displacements (several tens of kilometres) where the place of origin of the nappe is no longer identifiable. As we have already noted in the chapter devoted to the geometry, the displacement is the result of two components: rigid translation and internal strain. From a theoretical point of view, there is no particular relationship between the displacement achieved by internal strain and the displacement accomplished by rigid translation. Even before discussing the relationship between the direction of displacement and finite strain markers such as stretching lineation (e.g. Gamond 1980, Burg et al. 1987), it should be remembered that displacement related to internal strain can be absolutely negligible from the point of view of the quantity of displacement produced by rigid translation, and that the two directions can be very different.

In the case of the nappe of Laksefjord (already mentioned p. 21), only seven kilometres have left a mark, the internal strain, that the geologist can decipher in the field. The additional seventy kilometres are only an hypothesis established from paleogeographic reconstructions of the mountain chain before the thrusting (cf equally the example given by Boullier and Quenardel 1981). In experimental modelling of gravitational spreading, we have seen that in the upper part the displacement vectors were, at each moment, globally perpendicular to the stretching associated with the internal strain of the material. The surface grid shows a coaxial strain (the initial squares deform into rectangles) and rotational (the principal axes of strain turn during time in relation to an external marker). The displacement trajectory does not even correspond to the total translation vector but to the sum of the translation vectors at each moment of the experiment (Fig. 53).

Therefore, firstly it should be admitted, that the displacement produced through rigid translation, the importance of which has been seen in the case of shallow crustal nappes, may be neither determined (in direction) nor quantified through the methods of structural geology, and that this problem must be principally approached by the means of paleogeographic reconstructions and paleomagnetic studies. However some hypotheses are allowed, the limitations of which should be borne in mind, which sometimes allow the displacement trajectories to be proposed.

When deformation occurs through simple shear, for example at the base of an allochthonous unit along a layer of low mechanical strength, it is generally considered, particularly when the strain is significant and where the schistosity is sub-parallel to the plane of shear, that the stretching lineation relates to the displacement direction associated with the internal strain. Once again, to assume that this direction gives the direction of transport implies that one considers either that the rigid translation is non-existent or negligible, or that the rigid translation occurs in the same direction as the simple shear (which can be a reasonable hypothesis, particularly in nappes emplaced by ductile gravitational gliding). To add to simple shear a component of pure shear with the stretching parallel to the direction of shear naturally allows the same interpretation. Certain criteria such striae on bedding planes, interpreted as small intra-nappe replicas of rigid translation occurring at the base, can be compared to the direction of ductile stretching to validate the hypothesis.

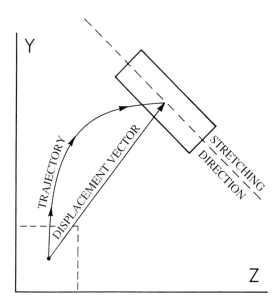

Figure 53 - Relationship between displacement trajectory, displacement vector and stretching of the material at the surface of experimental models of three dimensional spreading.

In any event, in the case where the stretching and the displacement can be related, the stretching lineation only allows the last direction of displacement suffered by the allochthonous unit to be estimated. An estimation of the changes in direction of the thrust can be made by the study of incremental strain markers. Amongst these, mineral fibres crystallised during the course of the deformation in veins and pressure shadows are well adapted to this problem and have been used several times with success (Merle and Brun 1984, Gourlay 1986, Dietrich and Durney 1986, Spencer 1989) (Fig. 54). Fibres crystallise in the direction of the incremental stretching and thus record the directional changes of stretching during time (Choukroune 1971, Durney and Ramsay 1973). The rotation of fibres in veins and pressure shadows (observed in the plane of schistosity) allows some hypotheses to be made about the changes of direction of displacement of the studied unit. There again, the interpretations must be moved forward cautiously and from several convergent arguments. The study of fibres can, for example, be coupled with the study of superimposed folds. Successive vergences must be consistent with the successive stretching directions (e.g. Townsend 1987). These parallels between deformation and displacement direction appear to be better adapted to the case of two dimensional gravitational gliding (where the slope produces an internal strain whose stretching direction is identical to the direction of the rigid translation).

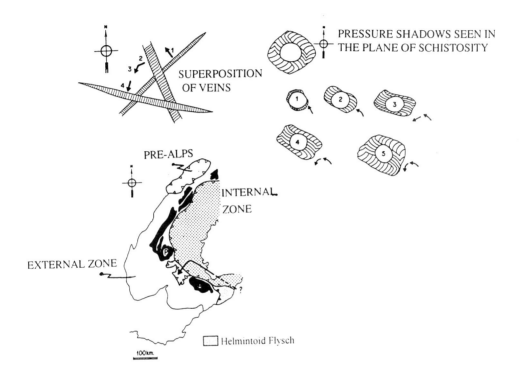

Figure 54 - Example of translation path reconstructed from markers of incremental strain. Above: the superposition of veins and the rotation of fibres in veins and in pressure shadows shows a systematic anticlockwise rotation of the incremental stretching. Below: trajectory assumed of the displacement of the flysch at Helminthoides in the western Alps (after Merle and Brun 1984).

4.4 Folding

4.4.1 ACTIVE FOLDING.

Folding is said to be *active* (Donath and Parker 1964) where it is the result of a mechanical instability, as is the case for the example where the beds of a stratified series have different rheological properties and because of this fact introduce contrasts in mechanical strength (Biot 1961, Ramberg 1961). Through the study of strain patterns it has been amply shown that one of the principal characteristics of strain in nappes is an increase of the strain, essentially the component γ of simple shear, towards the base of the thrust. In mylonite zones, in particular in the decollement zones, folds of all scales form (though shear) whose axes are initially perpendicular to the direction of shear. The intensity of strain reorientates the fold axes and all other linear markers, in the direction of stretching, which means in the direction of the shearing (Escher et al. 1975, Skjernaa 1980). This reorientation is often found at the base of overthrusts where sheath folds (cf Quinquis et al. 1978) are frequently found (Rhodes and Gayer 1977, Williams 1978). Thus a dispersion of the direction of the fold axes, between folds parallel to the direction of the stretching lineation (folds formed in the first stages of the deformation and totally reoriented) and those perpendicular to the lineation (non-reoriented folds formed in the final stages of the deformation) are found. This localised folding at the base are common to all allochthonous units whatever the emplacement mechanism.

Within the allochthonous unit, in models of rear compression folds are particularly well developed. They are interpreted to be the consequence of horizontal stress applied at the rear of the allochthonous unit. We have already indicated the shape in section of the axial planes which is overall defined by a concave shape from the summit towards the base, because of the gradient of the component γ whose consequences on the reorientation of the fold axes at the base of the allochthonous unit have been mentioned above (Fig. 55). In the upper part, where the component α is predominant, the stretching lineation is sub-vertical and therefore perpendicular to the fold axes (cf the example of the Appalachians thrust in South Virginia (U.S.A.), Reks and Gray 1983). Upright folds (associated with the component α and therefore with horizontal shortening) are more or less tight following their position in the vertical: the angle of opening between the two limbs regularly diminishes, from 30-50° at the top, to 20-30° at an intermediate level and less than 20° at the base (e.g. Gray and Willman 1991). In gravitational nappes, the component α (greater than 1) does not allow the formation of folds. However, the radial shortening in 3D spreading produces a fold belt whose axes are concentrically arranged. With the notable exception at the base where the component γ is predominant, the stretching lineation is equally concentric and therefore parallel to the fold axes. Again, because of the distribution of the two components α and γ, the axial planes of these folds are concave, steeply dipping at the surface and bedded towards the base.

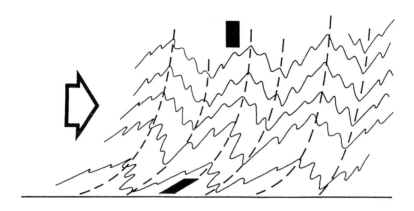

Figure 55 - Concave shape of the traces of axial planes of folds in a rear compression. The two black rectangles represent the internal strain at the top (pure shear associated with horizontal shortening) and at the base (combination of strain at the top with some simple shear) (modified after Gray and Willman 1991). Compare with Figure 45.

In thin-skinned thrust tectonics, the formation and/or the evolution of a thrust fault is associated with folding whose terminology has been given in the first chapter (cf for example the natural examples given by Boyer 1986). Classically, three interactions between fold and thrust fault are distinguished; *fault-bend fold*, *fault propagation fold* and *detachment fold* (Fig. 56). The fault propagation fold and the detachment fold are both associated with the termination of the thrust fault either on a ramp or on a flat (Fig. 56). The limit of a thrust fault (e.g. tip line, p. 14) is the place where its displacement progressively reduces to reach a zero value. The discontinuous displacement along the fault is therefore transformed into continuous internal strain, in front of the tip line (Williams and Chapman 1983). This internal strain globally relates to the horizontal shortening which causes the formation of

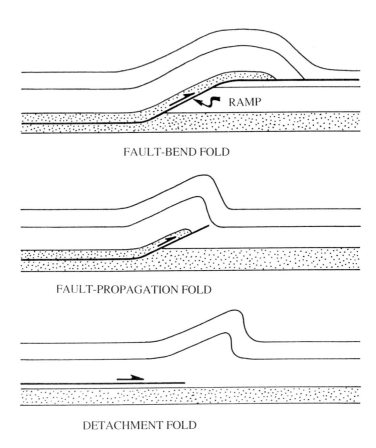

Figure 56 - The three classic interactions between fold and thrust faults (modified after Jamison 1987).

a fold, if the mechanical contrasts between the layers allow it, which is generally the case in stratified sequences. The fault-bend fold relates to another mechanism since it is formed immediately after the formation of the emergent ramp when the stratified sequence passes from the lower flat to the topographic surface. The formation of this fold is only the geometric adaptation of the sedimentary beds at the convex surface which forms the emergent ramp and the flat of the topographic surface. Therefore it is a passive fold like those which we will look at in the following section.

4.4.2 PASSIVE FOLDING.

A fold is called *passive* (Donath and Parker 1964) where it occurs in a medium of uniform rheology, without any mechanical contrast, either through purely kinematics amplification of an initial perturbation or by heterogeneous strain of initially planar markers. In fact, folding through buckling (i.e. active folding) is not the only possible type of folding in overthrusting, in particular in nappes emplaced by gravity. The study of gravitational spreading of glaciers is, from this point of view, very informative. In the glaciers, isoclinal folding of the stratification of the ice, in particular near to the basal contact is frequently observed (e.g. Chamberlin 1928, Hambrey 1977, Talbot 1979). Now, the different layers of stratification found in glaciers have practically the same rheological properties, which means that the folding occurs without viscosity contrast, by a purely kinematic process. The mathematical theory of this mode of folding as well as its application to natural examples has been carried out by Hudleston (1976, 1977). This model takes into account (i) irregularities of the basal surface and (ii) the vertical gradient of velocity. This gradient, calculated from the vertical velocity profile (cf Fig. 29), and which is common to all flows, indicates a reduction of the displacement velocity of the particles towards the base. It is associated with an increase of the γ component towards the base. At the base of a flow (such as it is), the stream lines are generally stable and related to the trajectories of the particles [22]. Following a modification of the boundary conditions (for a glacier, it can occur due to strong snowfalls), the stream lines will be affected and will recut the layering relating to more ancient particle trajectories. On the same interface (i.e. the boundary between two beds), the displacement velocities are no longer the same because of the vertical velocity gradient, and the most rapid points tend to catch up with the slowest points, causing folding (Fig. 57). The folds thus formed are folds of a similar type (Ramsay 1967, p. 367) and their asymmetry indicates without ambiguity the sense of displacement of the glacier. In the case of nappes, the conditions necessary for this type of kinematics folding could be achieved if the basal contact is an irregular

---

[22] A stream line is defined, at time t, by the curve whose constitutive points are displaced all at the same velocity. Velocity vectors of the same magnitude are therefore tangents to the corresponding stream line.

surface. The temporary deviation between the sedimentary layering and the stream lines, causing the folding, would again be associated with variations in the boundary conditions such as would occur by erosion, or by the arrival of a nappe from a more internal region on the surface above the deforming zone (Hudleston 1977). One such mechanism of folding has been proposed for certain kilometre scale folds localised in the hollow of the basal surface of the gravitational nappe of the Flysch at Helminthoides in the western Alps (Merle 1982).

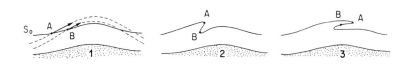

Figure 57 - Formation of a passive fold at the base of a flow above a relief of the basal surface (explanation in the text) (modified after Hudleston 1977).

Some laboratory experiments have been produced to analyse in more detail the geometry of folds produced in this manner (Brun and Merle 1988). The experiments have clearly shown the validity of Hudleston's theoretical model and have allowed the geometry of the folds produced in this way to be more clarified. In particular, the degree of cylindricality of the fold has been appreciated to be a function of the cylindricality of the irregularities arranged at the base of the models. The results go beyond the knowledge of passive folds and may also be applied to active folds (Brun and Merle 1988). Above all the following points may be summarised:
1. The cylindricality of folds is directly related to the cylindricality of the irregularities which cause the folding: non-cylindrical obstacles give non-cylindrical folds.
2. If the deformation is intense, the non-cylindrical folds evolve as sheath folds whatever the strain regime (coaxial or non-coaxial).
3. The size of the non-cylindrical folds or sheath folds is equal to the dimension of the irregularities of the basal surface.
4. The viscosity contrasts between the layers introduces some mechanical effects which cause refolding of the folds formed by kinematics amplification. Therefore care should be taken in the field of making hasty interpretations in terms of polyphase deformation.

4.5 Emergent Ramp

The type of internal strain associated with the passage of a stratified series from a lower flat to a structurally higher flat through the intermediary of a ramp is highly debated and has not yet received a general model recognised by everybody. During its propagation along the initial flat, the limit of the frontal extension of the thrust unit is accompanied by an internal strain characterised by a horizontal shortening (e.g. Williams and Chapman 1983, Marshak and Endelger 1985, Nickelsen 1986) with produces local thickening. At first, it can be considered that the passage of the stratified series from the lower flat to the upper flat occurs (i) by two successive vertical shears which are of opposite sense [Mode A] (Sanderson 1982), (ii) by flexure [Mode B] (Sanderson 1982, Fisher and Coward 1982, Suppe 1983, Gray and Willman 1991) and (iii) by compression sub-parallel to the displacement [Mode C] (Beutner et al. 1988, Geiser 1988).

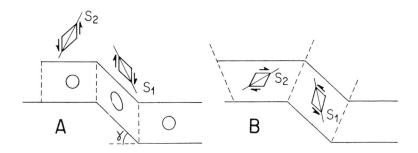

Figure 58 - Internal strain during the passage on a ramp (modified after Sanderson 1982). Left: the deformation occurs through two successive shears which are of opposite sense. Right: the deformation occurs through flexure of the layers

Mode A. In this first model, the fault-bend fold is produced by vertical shear. These shears have an opposite sense on the ramp and on the upper flat (Fig. 58). The principal direction of the internal strain associated with these two simple shears are clearly distinct. This results in the schistosity S_1 formed on the ramp being cross-cut by a schistosity S_2 of opposite dip on the upper flat. The layers carried onto the upper flat should in this case possess two superimposed schistosities whose dip is directly dependent on the dip of the ramp itself (the dip of the ramp gives the values of the two γ of opposite sense suffered by the layers). In this model the two shears of opposite sense cancel each other out mutually, so that the layers reduces in thickness on the ramp then recover their initial thickness on the upper flat. If the successive schistosities record the progressive history of the deformation, the total deformation is nothing. The promoter of this model himself judges it to be not very realistic from the geological point of view (Sanderson 1982).

Mode B. In this second model, the rotation of the layers on the ramp then on the upper flat is accommodated by a simple shear parallel to the sedimentary layers (Fig. 58). The type of deformation is analogous to the one which is observed in folds deforming by flexure (flexural-flow fold, Ramsay 1967, p. 392), the best image of which is one of the folding of a packet of cards where a reversal of the relative displacement of the cards each in relation to the others on either side of the axial plane is found. The thickness of the layers is therefore conserved on the ramp. Again, the shear sense becomes reversed on the upper flat and the progressive deformation of the beds should allow the appearance of two superimposed schistosities of opposite dip. The objection to this model is not really to consider the progressive history of the passage of the layers from the ramp to the upper flat but to propose a model of folding of the fault-bend fold in its final state, then to artificially superimpose strain states of each of the limbs. However, some field studies show that simple shear parallel to the layers is one of the deformation mechanisms present within ramps (e.g. Gray and Willman 1991).

Mode C. This third model takes into account the resistance met by the overthrusting unit at the approach to the ramp, whose opposite inclination requires a supplementary energy to overcome this obstacle. This resistance to movement is overcome by an internal strain which locally produces, on the ramp, a shortening and a thickening of the layers causing the formation of a schistosity sub-perpendicular to the bedding plane (Fig. 59). This shortening associated with resistance to movement on the ramp is compatible with the orientation of the calculated stresses on the ramp by some authors (e.g. Wiltschko 1981). The striking characteristic of this model is that the dip of the schistosity is therefore opposite to the one of the ramp, which sometimes is actually observed in the field (e.g. Beutner et al. 1988, Geiser 1988). Data resulting from field studies being relatively contradictory, it can be supposed that the three models become combined to give some hybrid models (cf Knipe 1985 by way of example).

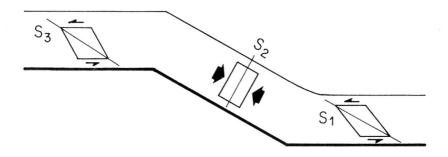

Figure 59 - Passage of a ramp by shortening parallel to the direction of displacement.

Equally it can be envisaged that the internal strain is localised in a decollement level and that the stratified series situated above displaces itself on the ramp by rigid translation without internal strain (Taboada et al. 1990) (Fig. 60). The internal strain in the decollement level along the lower flat is related to simple shear, in accordance with the terms of Kelhe's mechanical model. At the level of the ramp, the internal strain in the ductile level combines simple shear and pure shear (Taboada et al. 1990). The α component corresponds to a stretching in the direction of the displacement. This combination of the two components α and γ can be shown by the geometries of pressure shadows within the ramp (Malavieille and Ritz 1989). This deformation produces a progressive thinning of the ductile layer along the ramp, then its disappearance on the upper flat (Fig. 60). The disappearance of the decollement level in front of the allochthonous unit can therefore causes an increase in the resistance to movement which could cause the creation of the new ramp behind the preceding ramp where the decollement level is still present. Experiments on scaled models have been conducted in the laboratory in an attempt to better understand the kinematics of the deformation associated with the passage of a stratified sequence over a ramp. The displacement on the ramp produces first a conjugate thrust, which terminates at depth at the intersection between the ramp and the flat. A second, then a third conjugate (and like this in

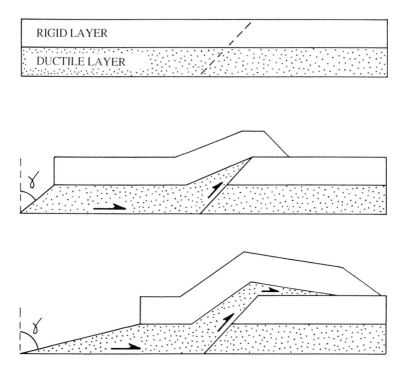

Figure 60 - Passage of a ramp by the internal strain of a basal layer of low mechanical strength and rigid translation of the upper part (modified after Taboada et al. 1990).

succession) form behind the first during displacement on the ramp. Analysis of the kinematics of the experiments shows that the first conjugate is transported on the ramp until the time when it ceases to function at which point a new conjugate forms behind, at the same place where the preceding was formed (Morse 1997, Malavieille 1984, Bale 1986, Ballard et al. 1987, Tondji Biyo 1993, Merle and Abidi 1995). Between these conjugate faults, the layers remain close to the horizontal. However, these discontinuous displacements allow the average trace of the layering above the ramp to become progressively parallel to the ramp itself (Figure 61).

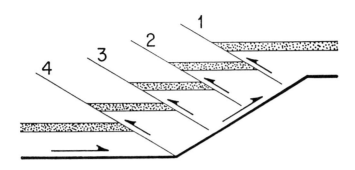

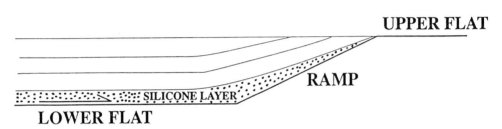

Figure 61 - Top: a conjugate set of shear planes (backthrusts) allows average trace of layering to become parallel to the ramp. Numbers correspond to the order of appearance of conjugate planes (drawn from experiments with sand layers, no erosion). Below: with erosion along the upper flat and a decollement horizon along the lower flat, the layering is rotated into parallelism with the frontal ramp and the thrusting is active without any backthrust until the end of the experiment.

The number of conjugate faults is dependent on the brittle/ductile ratio (= ratio in the allochthonous unit between the thickness of the ductile layer at the base of the flat and the brittle layers which override it), which reduce in number with the increase in thickness of the ductile layer. No conjugate faults form where the ductile layer reaches a thickness equal to the brittle part (Merle and Abidi 1995). In experiments with continuous erosion of the material involved above the ramp, only a single conjugate is formed and the functioning of the ramp is prolonged indefinitely. Brought to the surface, the conjugate is therefore eroded in its turn and the final geometry relates to a single monocline structure resulting from the flexure of the stratified sequence above the ramp (Merle and Abidi 1995) (Figure 61). In this final case, the conditions close to the model of Figure 60 are found again, and the brittle layers of the stratified sequence are parallel to the ramp owing to the ductile deformation of the underlying ductile layer.

Finally, in 1983 John Suppe proposed a purely geometric model which has engendered considerable enthusiasm and led to a multitude of geometrical developments and refinements of thrust structures in ramps and flats (e.g. Mitra 1986, Jamison 1987, Endignoux and Mugnier 1990, Zoetemeijer and Sassi 1992). In this model, the crossing over of the ramp is accompanied in the beginning by the formation of a 'kink band' at the base of the ramp (Figure 62). The upper axial plane then migrates along the ramp, at the same time as the stratified sequence, while the lower axial plane remains fixed to the base of the ramp. This model, geometrically very elegant, supplies a real and complete rotation of the beds until total parallelism with the ramp is achieved (Figure 62). However, there are two geometric hypotheses to respect for the model to be viable mathematically: the thickness of layers remains constant during the displacement and the dip of the ramp must be less of equal to 30 degrees.

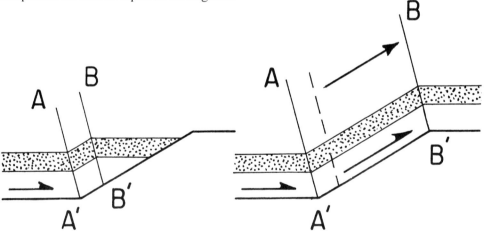

Figure 62 - Suppe's model: riding over the ramp is achieved by the formation of a kink band (AA' - BB') followed by upward migration of the upper axial surface (BB').

4.6 Lateral Ramps

4.6.1 THE BOW AND ARROW RULE.

The majority of authors considers that a thrust fault initially forms through the rupture of the rock material at a point, and that it propagates at the same time both towards the front and laterally (Douglas 1958, p.130). This results in numerous thrust units (or thrust belts, e.g. the Jura) having an arcuate shape: it is indicative of a more significant amount of displacement at the center of the thrust unit than towards the two lateral boundaries. Ideally, the amount of displacement reduces symmetrically from the centre towards the two lateral borders until it reaches a zero value. The maximum quantity of displacement is therefore measured on the bisector of the line joining the two ends of the thrust unit (Fig. 63) and it is

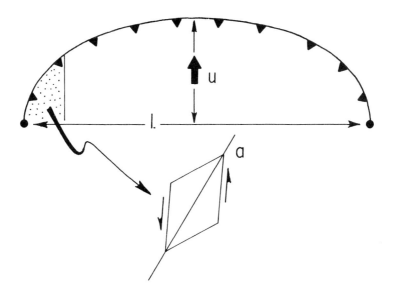

Figure 63 - The bow and arrow rule for measuring the amount of displacement (U) (modified after Elliott 1976). Within the lateral borders, the transcurrent shear (i.e. vertical strike-slip) produces folds whose axes (a) are oblique in relation to the displacement direction.

generally stated that the arcuate form and the amount of displacement are related by a simple relationship (Elliott 1976). The quantity of displacement is close to 7% of the length of the line joining the two ends of the thrust unit. This bow and arrow rule (Elliott 1976b) assumes that the tip line (cf p. 14) of the thrust is clearly identified, in particular laterally. In effect it is known that numerous thrust faults are laterally rooted in folds. These folds are often asymmetric and disappear equally, according to a characteristic distance of eight kilometres, independent of the size of the overthrust (Elliott 1976b). These folds are the expression of the lateral propagation of the thrust fault which requires a certain amount of ductile deformation before the rupture may occur.

The bow and arrow rule implies that the thrust fault maintains its continuity from the centre towards the lateral borders. This differential displacement is produced by a certain amount of transcurrent shear (i.e. whose shear planes are vertical). In shallow crustal overthrusts, the lateral continuity is maintained so that the angle of shear does not exceed 35 - 40°. Beyond that, lateral continuity is broken by the appearance of strike-slip faults which are more or less parallel to the displacement direction and/or by the formation of imbricated thrusts (Wilkerson 1992).

On a map, the tip line of the overthrust is bent at the level of the two lateral borders to become sub-parallel to the direction of displacement itself. In the two border zones, the transcurrent shear becomes a determinant element of the internal strain which adds itself to the characteristics of the deformation within the thrust unit. To measure this influence, it is enough to remember that transcurrent shear produces folds whose axes are, since their formation, parallel to the stretching lineation and oriented at 45° to the shear planes. This particular fold geometry during a transcurrent shear is easily confirmed by experiments on laboratory models (Odonne and Vialon 1983) or through numerical simulations (Ridley and Casey 1989). Thus, at the level of the lateral borders, folds with axes oblique in relation to the displacement direction should be expected (Fig. 63). If the transcurrent shear intensifies, these fold axes turn with the stretching lineation and become parallel to the displacement direction. The internal strain in these lateral zones is therefore specific and deserves to be examined in more detail.

4.6.2 THE WRENCHING COMPONENT.

To determine the internal strain within the lateral borders means integrating this transcurrent component γ_T with the previously described components, namely the

component of simple shear acting tangentially to the sedimentary layers (which we will now name γ_c) and the component of pure shear α producing either a thickening ($\alpha<1$) or a vertical thinning ($\alpha>1$). Three cases can be envisaged:
1. Constant thickness ($\alpha=1$): the two components of simple shear γ_c and γ_T determine themselves alone the characteristics of the internal strain (e.g. Brun 1977, Sanderson 1982). This deformation can relate to the one observed in the lateral borders of a ductile gliding.
2. Vertical thickening ($\alpha<1$): the component γ_T is combined with the component γ_C and with the component α to take into account the deformation occurring in the lateral borders of a rear compression or an extrusion (e.g. Sanderson 1982, Coward and Kim 1981).
3. Vertical thinning ($\alpha>1$): the component γ_T is combined with γ_C and α, which can account, for example, for the internal strain in the lateral ramps of spreading or of gliding-spreading.

Although these different combinations produce three-dimensional strain (in the sense where the intermediate axis λ_2 can be different to 1), it is important to indicate that the particle trajectories within the lateral borders remain parallel to the ones in the central zone. In other words, these models exclude all radial or divergent types of deformation and relate to the simple case of flow in a channel where the two lateral borders are parallel. For each of the cases, the precise characteristics of the internal strain depends on the relative proportions of the two or three components implicated in the calculation. By way of example, in the case where α is equal to 1, the position of the fold axes (that we have mentioned when γ_T is the only active component) depends on the relative proportion of γ_C and of γ_T (Coward and Potts 1983). The fold axes appear between forty five and ninety degrees to the displacement direction (Fig. 64). With an increase in total shear, the fold axes are progressively reoriented in the direction of transport but the fold axes remain straight. The fold asymmetry, given by the component γ_C, is therefore conserved so that the imposed rotation produces a sense of (apparent) overturning towards the lateral border (Fig. 64) (Coward and Potts 1983).

Still in the case of a constant vertical thickness ($\alpha=1$), the position of the three principal axes of the strain ellipsoid during time can be followed on a stereogram (Fig. 65). Again, it is stated that the position of the three axes depends on the relative proportions of γ_C and γ_T, with the exception of the λ_2 axis constant in orientation during time for a given ratio (Brun 1977, Sanderson 1982). In spite of these different initial positions, the rotation of the stretching axis λ_1 and of the shortening axis λ_3 with an increase in the deformation is comparable, and tends

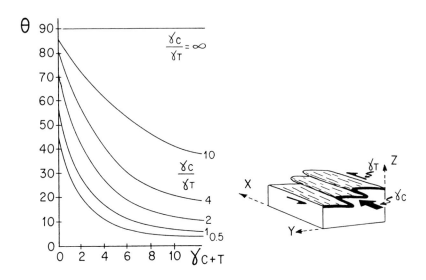

Figure 64 - Orientation and reorientation of fold axes according to the relative proportion of the two shear components γ_C and γ_T (after Coward and Potts 1983).

towards parallelism between λ_1 and the direction of displacement. It is equally interesting to follow the direction and the dip of the schistosity as a function of the increase in the component γ_T for a γ_C constant, which relates to the evolution of the schistosity plane from the centre of the thrust towards its lateral border. A verticalisation and a rotation of this plane which becomes oriented parallel to the direction of the lateral ramp is seen (Fig. 65).

This variation in the position of the plane of schistosity towards the lateral ramp is equally true if the component α is varied. However, the following differences should be kept in mind. In the case of the rear compression ($\alpha < 1$), towards the top of the allochthonous unit, which means where γ_C is negligible, the plane of schistosity already close to the vertical at the centre of the unit becomes parallel to the lateral border through a rotation around a vertical axis. On the other hand, towards the base of the allochthonous unit, where the component γ_C is high and therefore the schistosity close to the horizontal, the passage to the lateral border is accompanied by a verticalisation of the schistosity (Sanderson 1982). Likewise, in the case of two-dimensional gravitational spreading ($\alpha > 1$), in particular towards the top of the allochthonous unit, which means where the component γ_C is weak or non-existent, the sub-horizontal plane of schistosity becomes vertical towards the lateral ramp while conserving the stretching axis λ_1 close to the horizontal. In all the cases, within the lateral border, the stretching axis λ_1 takes an oblique direction

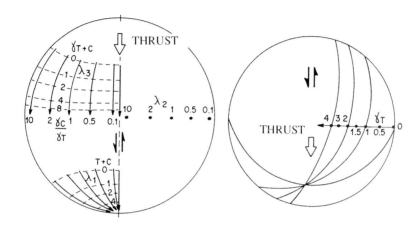

Figure 65 - Orientation and evolution of the principal strain axes (left) and of schistosity (right) as a function of the two shear components γ_c and γ_T (modified after Sanderson 1982).

in relation to the particle trajectories (which are parallel to the direction of the channel) but tends to become parallel to these trajectories with an increase in deformation.

One of the principal characteristics of internal strain in these lateral borders is the shape of the strain ellipsoid which closely depends on the component of pure shear α (Sanderson 1982). Thus, above all the following results can be deduced:
1. The ellipsoid is plane ($\lambda_2 = 1$) where the component α is equal to 1.
2. The ellipsoid is of flattening type ($\lambda_2 > 1$) where the component α is less than 1.
3. The ellipsoid is of constriction type ($\lambda_2 < 1$) where the component α is greater than 1.

Again, it proves to be that the analysis of the internal strain in the allochthonous unit, but this time within the lateral borders, is one of the best arguments, if it is not the best, to contrast sharply between different models of emplacement of nappes such as 2D ductile gliding ($\alpha=1$), 2D rear compression ($\alpha<1$) or 2D spreading sensu lato ($\alpha>1$).

CHAPTER 5

5 A FEW EXAMPLES FROM THE ALPINE CHAIN

Although in passing a certain number of field examples have already been mentioned, it is useful to replace nappes and overthrusts back into their natural environment and position them in the structural context of a mountain chain. We will take the example of the Alpine chain where the majority of the emplacement mechanisms introduced during the preceding chapters can be identified. The position of nappes and overthrusts in the Alps is characteristic enough so that this chain can be considered to be an ideal case, at least from this point of view... In reviewing, very succinctly, the major zones of this chain (i.e. the internal zone, the Helvetic zone, the Pre-Alps, the mollassic basin, the Jura and the southern-Alpine zone), it will be seen that the different mechanisms described until now become integrated in the zoneography of the chain which results from its history through time.

5.1 The Internal Zone: Crustal Stacking and Basement Nappes

The Alpine chain results from the collision between two tectonic plates: the European plate situated in a northern position and the African plate (or more precisely its Adriatic headland) situated in a southern position (Fig. 66). The history of this collision stretches over nearly a hundred million years, which makes the kinematics reconstruction of the chain particularly difficult and hypothetical. To achieve this, it is sufficient to consider the large number of

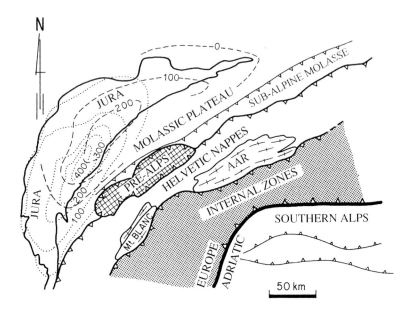

Figure 66 - Simplified structural outline of the central Alps. The deformation in the Adriatic plate is weak compared to that observed in the European plate. The internal zone of the European plate is made up of intensely deformed and metamorphosed sub-horizontal basement nappes. The external zone where the vertical foliations of the basement massifs of Mont Blanc and of the Aar are exposed relate to the zone of extrusion of Helvetic material. Further to the North, nappes of the Pre-Alps which come from a southern origin, rest on the molassic basin deposited during the orogeny. The last event of the compressive episode in the Alps, the Jura, has formed uniquely where the decollement level of the Triassic saliferous beds are found, whose isopaques (in meters) are indicated by the dotted and dashed lines (dotted: Keuper, dashed: Muschelkalk).

models proposed for the evolution of the internal zone (e.g. Dal Piaz et al. 1972, Caby et al. 1978, Platt 1986, Gillet et al. 1986, Steck 1987, Hsu 1991, Merle and Ballevre 1992). The experimental study of the tectonic evolution of the chain gives place equally to highly contrasting approaches (e.g. Malavieille 1984, Merle and Guillier 1989). However, it is known with certainty that the Adriatic plate in its movement towards the North has overthrusted the European plate thus creating a considerable thickening within the internal zones of the chain. This crustal overthrusting, easily identifiable in the field where enormous tectonic outliers of the Adriatic plate still remain (i.e. the Dent Blanche klippe), has progressively become inactive and the compression has been carried through to the European plate where other crustal overthrusts, visible mainly owing to geophysical methods, have been formed. The classic process is the migration of crustal overthrusts towards progressively more external zones, a phenomenon to which we will return. The current image of these crustal overthrusts, such as they can be interpreted in the light of geophysical data, is apparent in Figure 6.

In the internal zones of chains where crustal stacking occurs the deformation is remarkably intense (cf Fig. 5); this results from absolutely exceptional pressure and temperature conditions. We have indicated in the first chapter that the emplacement mechanisms of basement nappes (Fig. 67) which would become individualised under such conditions were not identifiable, in particular because these nappes are too dependent on the generally complex tectono-metamorphic evolution of the chain. By way of example it should be noted, that some of these nappes have experienced highly contrasting evolutions sometimes going to an eclogite metamorphism with the formation of coesite in the first stages of the collision (burial greater than 70 kilometres) then to greenschist facies in the final stages of their tectonic history (burial close to 15 kilometres). Studying these basement nappes returns us to the study of the history of the orogeny itself and wanders from the subject treated in this book.

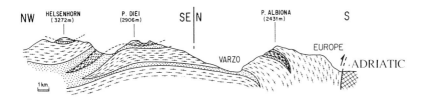

Figure 67 - The internal zone relates to a stacking of pre-Mesozoic basement nappes (crosses) bounded by fine bands of Mesozoic sedimentary material (dotted).

5.2 The Helvetic Nappes: Extrusion, Spreading and Ductile Gliding

The Helvetic nappes are situated at the northern boundary of the internal zone (Fig. 68). This position, at the border of the most deformed zones of the chain, explains their major structural characteristics. This comprises predominantly a stacking of nappes made up of sedimentary material whose lower unit, the famous Morcles fold-nappe, has been mentioned in the models of extruding-spreading (cf Fig. 52). The displacements experienced by these nappes are relatively moderate. Following the compression exerted by the internal zones and the intense shortening of their underlying basement, (i.e. the crystalline massifs of Mont Blanc and of Aiguilles Rouges), the Mesozoic cover sequence has been extruded from its original basin to spread and/or glide ductily towards the foreland. We will not return to their characteristic internal strain, introduced in the chapter devoted to kinematics. Let us say only that this Helvetic structure can be described as the result of a huge backpush of the sedimentary series, resulting from intense horizontal shortening of the backward internal domain. If the horizontal compression is the first cause of this backpush, gravity becomes active as the difference in altitude with the foreland becomes too significant. Therefore, in a combined way, gliding and spreading occurs. Ductile gliding, in particular, is facilitated by certain levels of low mechanical strength such as the Triassic evaporites or the Middle Jurassic and Cretaceous schists.

Metamorphic conditions at the time of the deformation are not negligeable and reveal temperatures which fluctuate between 200°C and 400°C for depths which hardly exceed ten kilometres for the most deep units.

5.3 The Pre-Alps: Rigid Gliding

In mentioning the sort of sedimentary cover sequences located at the limit of the internal and external zones, we have forgotten sedimentary cover sequences initially situated within the internal zones in a shallow crustal position. These sequences, following crustal overthrust faulting, are most often buried to great depths. However, some of these sedimentary sequences, rapidly uplifted to considerable altitudes by the shortening which affects the internal zones, sometimes escape this burial by gliding rigidly towards the foreland. This generalised gliding of the most shallow crustal sedimentary sequences carries them as far as the limit between the external and internal zones where they pass over the nappes which we have just mentioned (i.e. the Helvetic nappes).

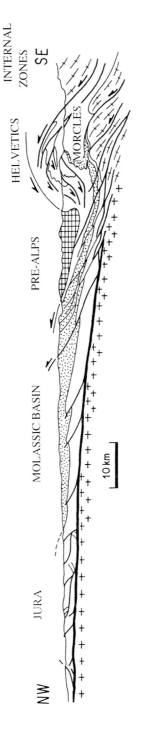

Figure 68 - Schematic cross section of the external zone of the Alps. The Helvetic nappes result from the extruding-spreading of sedimentary material located at the boundary between the internal and external zones. The Pre-Alps relate to a sequence of nappes which have slid rigidly over the Helvetic zone. The molassic basin which was deposited at the beginning of the Tertiary; due to crustal stacking in the internal zones, is affected by the compression towards the end of the Tertiary. The Jura is the ultimate evidence, both in time and space, of the migration of the deformation towards the foreland (after Homewood et al. 1988).

This is the case of the Pre-Alps of the Alpine chain (Fig. 68). Now located in front of the Helvetic nappes, the majority of its units comes from more internal zones from where they have slid without deformation (e.g. Plancherel 1979). There again, the Pre-Alps consists of a stacking of nappes, whose superposition reflects the progression of the global shortening towards the exterior of the chain. The units from the most internal zones are in effect situated at the top of the structure whereas the lower units come from less distant zones. The displacement associated with this rigid gliding is clearly more significant than the displacement associated with the Helvetic nappes but varies, of course, from one unit to another, the upper units having suffered the greatest displacements.

Rigid gliding starts before the deformation of the Helvetic zone but its passage over the Helvetics is on the whole contemporaneous with the formation of the Helvetic nappes, because of the progression of the deformation towards the exterior of the chain. On the other hand, the gliding of the Pre-Alps ends when these nappes reach the depression situated in front of the Helvetic zone, whilst the emplacement of the Helvetic nappes is not totally achieved. The basal contact of the Pre-Alps is therefore folded by the movements of the underlying Helvetic nappes, and of the molasses whose geodynamic importance will be mentioned in the following section. Some deformation by gliding and strike-slip faults therefore affects this edifice by complicating its whole structure (e.g. Plancherel 1979).

The sedimentary material of the Pre-Alps has not been buried before its gliding towards the foreland and reflects practically no metamorphism. It is generally unharmed of internal strain during its emplacement (i.e. rigid gliding), even if locally some particularly deformable sedimentary layers introduce a weak schistosity.

5.4 Molasses: Testimony to Crustal Thickening

Molasses are detrital sediments which are deposited at the front of the most external crustal overthrusts. They are the erosional product of the adjacent reliefs and are synchronous with the formation of the chain. This *foreland basin* (Dickinson 1974) is common to all mountain chains and results from lithospheric flexure under load due to crustal thickening in the internal zones (Fig. 69). Because of this accommodation phenomenon of crustal stacking, the subsidence rate is extremely rapid, which clearly differentiates these from subsidence basins observed in other tectonic contexts (Allen et al. 1986). The progression of crustal overthrusts towards the foreland through time makes this *flexural basin* an excellent marker of the tectonic evolution of the chain.

In the Alps, sediments of the molassic basin are primarily marine and deposit within the Helvetic region (North-Helvetic flysch). The progression of crustal overthrusts towards the foreland drags along the migration of the basin itself towards the most external zones. This migration is obvious by the progressively more northern position of the depocentres (Homewood et al. 1986). The most internal molasses are at the same time deformed and imbricated, so that classically the molasses can be subdivided into two zones: the sub-Alpine molasse in an internal position (deformed by the imbrication) and the molassic Plateau which is overall unaffected by this deformation. It goes without saying that the exact age of the sediments in this foreland basin allows the date of the deformation in the external zones as well as the rate of the migration of the deformation towards the North to be determined with great precision. The analysis of interbedded olistoliths in the marine molasse which are derived from the most internal nappes equally allows the emplacement of these nappes to be fairly precisely dated. Therefore it is the entire Tertiary history of the chain that the molassic basin of the Alps records with the degree of precision of our stratigraphic knowledge.

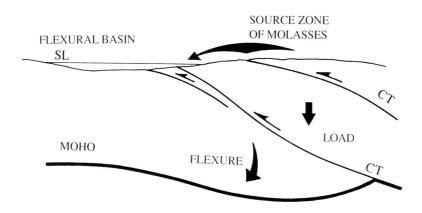

Figure 69 - Crustal overthrusts (CT) cause crustal thickening whose load causes a lithospheric flexure responsible for the formation of a foreland basin. The molasses which deposit in this basin come from the erosion of the adjacent relief. SL: sea level (modified after Homewood and Lateltin 1988).

5.5 The Jura: A Fold and Thrust Belt

The Jura is an arcuate chain situated at the front of the Alpine chain. Of very late stage formation in the general history of the chain, it relates to the last compressive movements recorded less than ten million years ago. Moreover according to some authors, the displacements in the Jura should not be totally over today (Naef et al. 1985). The displacement is associated with two mechanisms: rear push for the internal Jura and gravitational gliding for the external Jura (Mugnier and Vialon 1986).

The Jura relates above all to displacement of the sedimentary cover sequence over the undeformed basement rock. All the characteristics of a fold and thrust belt can be described, in particular as (i) a wedge-shape form showing an increase in thickness towards the rear and as (ii) a decollement level situated just above the undeformed basement (Fig. 70). By applying the bow and arrow rule, the arcuate shape of this chain can be used to indicate a differential displacement between the centre and the two lateral borders. This hypothesis is confirmed by the spatial arrangement of the isopaques (i.e. curves of equal thickness of a sedimentary level) of the decollement level at the base of the belt. In effect, it is stated that these isopaques closely model the arcuate shape of the Jura (Rigassi 1977) (Fig. 66). This indicates that the displacement is only produced at the place where the salt-bearing Triassic was present by significantly reducing the friction at the base of the belt. Laterally, the displacement reduced progressively until it dies out where the decollement level disappears. In this sense, the Jura represents without any doubt one of the most outstanding examples of the role of the decollement level in the formation of fold and thrust belts.

Equally it is very interesting to point out that this decollement level plays an important role in the deformation of the adjacent molasse. Again, the saliferous Triassic is only present behind the Jura and disappears further to the East. This distribution of the saliferous Triassic level allows the hypothesis of a general gliding of the Western molassic Plateau on its basement to be proposed (Homewood et al. 1986), which would explain the rear push experienced by the Jura. Again, the absence of the decollement level in the East, which prevents the molassic Plateau from gliding as a whole mass, results in an increase in tectonic imbrication at the front of the Helvetic nappes. Equally, a reasonable explanation is found for the widening of the sub-Alpine molasse zone towards the East, that is where the compression may not be converted into gliding of the molasses and of the Jura on the saliferous layer, and where deformation occurs by imbricate overthrusting immediately at the front of the Alps strictly speaking (cf Burkhard 1990).

A Few Examples from the Alpine Chain

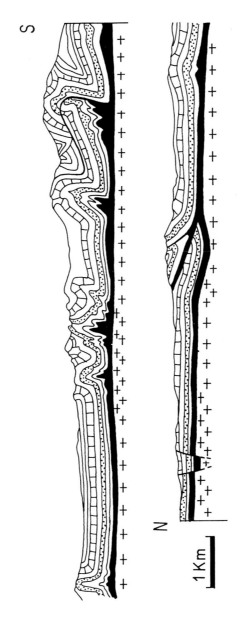

Figure 70 - Cross-section of the Jura showing the decollement of the sedimentary succession on the undeformed basement (crosses) owing to the ductile saliferous Trias (black). Structure of folds and overthrusts (modified after Buxtorf 1916).

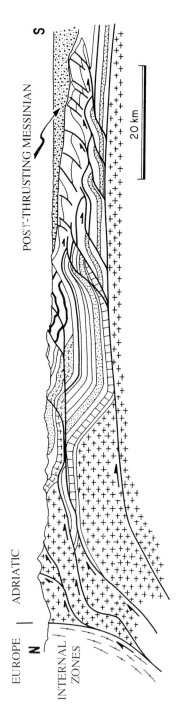

Figure 71 - Simplified cross-section of the southern Alps. The collision of the two plates causes backthrusting of basement blocks of the Adriatic plate which compresses the rear of the sedimentary sequences and develops a system of ramps and flats above the most external basement which remains unaffected by deformation (modified after Schonborn 1992).

Furthermore, in the chapter on the kinematics we have already indicated the strain partitioning resulting from the presence of this decollement level, as well as the role of this strain partitioning in the fold geometry of the Jura.

5.6 The Southern Alps: Rear Compression

To the south of the internal zone, in the Adriatic plate, the collision between the two plates has displaced the basement compartments which have uplifted and compressed the sedimentary cover rocks. The scale of these movements is limited but their interest lies in the sense of displacement which is oriented towards the South, which means in the opposite direction to the displacements which are generally observed in the Alpine chain, and which follows from the vergence of the large crustal stacking (Fig. 6). Such displacements towards the South are above all the mark of on-going convergence between the two plates after the principal collision. This late stage tightening thus gives the Alps of a fan-shape, even if this is clearly asymmetric with a predominance of the displacements towards the North. This deformation, which was produced in several stages until a very recent period, are very representative of rear compression, with a reduction in the intensity of deformation and structural complication with increasing distance from the overthrusted basement blocks in the direction of the Po plain.

The style of the deformation is very characteristic of thin skinned thrusting. A succession of ramps and flats are found which are directly related to rheological contrasts between different beds of the Trias (Schonborn 1992) (Fig. 71).

This concise overview of the different emplacement mechanisms of nappes and overthrusts found in the Alpine chain, although rapid, above all shows the importance of mechanical and kinematics studies of these geological structures in the understanding of the history of a mountain chain. The range of possible interpretations are first order data which all mechanical and/or kinematics models of the general evolution of the chain must (or should) take into account.

References

ALLEN, P.A., HOMEWOOD, P. and GRAHAM, G.D. - Foreland basin: an introduction. *Spec. Publ. int. Ass. Sediment.* **8**, 3-12 (1986).
APRAHAMIAN, J. and PAIRIS, J.L. - Very low grade metamorphism with a reverse gradient induced by an overthrust in Haute-Savoie (France), in *Thrust and nappe tectonics*, Geological Society Special Publ. 9, 159 - 165 (1981).
ARGAND, E. - Les nappes de recouvrement des Alpes Penniques et leur prolongements structuraux. *Beitrage zur Geol. Karte Schweiz*, **31**, 1 - 25. (1911).
ARGAND, E. - Sur l'arc des Alpes Occidentales. *Eclogae Geol. Helv.* **14**, 145 - 204 (1916).
ARGAND, E. - La tectonique de l'Asie. *Congr. Int.*, 13e Sess. 1/5, 171 - 372 (1924).
ARTHAUD, F. - Un exemple de relations entre l'étirement dans B, la dispersion des lineations et la courbure des axes de plis: la virgation des plis couches hercyniens du versant sud et de la Montagne Noire. *Rev. Geogr. Phys. Geol. Dyn.* XI, **5**, 523 - 531 (1969).
ARTHAUD, F. - Etude tectonique et microtectonique comparée de deux domaines hercyniens: les nappes de la Montagne Noire (France) et l'anticlinorium de L'Iglesiente (Sardaigne). *Publ USTELA* 1, Montpellier, 175, (1970).
AUBERT, D. - Le Jura et la tectonique d'écoulement; *Bull. Lab. Geol. Univ. Lausanne* 83, (1945).
AYRTON, S. - High fluid pressure, isothermal surfaces and the initiation of nappe movement. *Geology* **8**, 172 - 174 (1980).
BADOUX, J. - Tectonique de la nappe de Morcles entre Rhône et Lizerne. *Mat. Carte geol. Suisse* 143 (1972).
BAILEY, E.B. - *Tectonic essays, mainly Alpine.* Oxford Universiy Press, London, 200p. (1935).
BALE, P. - *Tectonique cadomienne en Bretagne nord, interaction décrochement chevauchement: champ de déformation et modélisations expérimentales.* Thèse Université Rennes, 361p. (1986).
BALLARD, J.F., BRUN, J.P., VAN DEN DRIESSCHE, J and ALLEMAND, P. - Propagation des chevauchements au-dessus des zones de décollement: modelés expérimentaux. *C. R. Acad. Sci. Paris* **305**, II, 1249 - 1253 (1987).
BALLY, A.W., GORDY, P.L. and STEWART, G.A. - Structure, seismic data and orogenic evolution of southern Canadian Rocky Mountains. *Bull. Can. Pet. Geol.* **14**, 337 - 381 (1978).
BAYLY, B. - *Mechanics in structural geology.* Springer Verlag. (1992).

BEACH, A. - Strain analysis in a cover thrust zone, external French Alps. *Tectonophysics* **88**: 33-346 (1982).
BEAUMONT, C., FULLSACK, P. and HAMILTON, J. - Erosional control of active compressional orogens. In *Thrust Tectonics* ed. McClay K.R., Chapman and Hall, 1 - 18 (1992).
BELOUSSOV, V.V. - *Basic problem in geotectonics*. McGraw-Hill ed., New York (1962).
BERTRAND, M. - Rapport de structure des Alpes de Glaris et du bassin Houiller du Nord. *Bull. Soc. Geol. France* **12**, 318 - 330 (1884).
BERTRAND, M. - Coupe de la chaîne de la Sainte-Beaume (Provence*). Bull. Soc. Geol. France* **13**, 115 - 131 (1884)
BERTRAND, M. - Ilot triasique du Beausset (Var). Analogie avec le bassin houiller Franco-Belge et avec les Alpes de Glaris. *Bull. Soc. Geol. France* **15**, 667 - 246 (1887).
BERTRAND, M. - Plis couches de la région de Draguignan. *Bull. Soc. Geol. France* **17**, 234 - 246 (1889).
BEUTNER, E.C., FISHER, D.M. and KIRKPATRICK, J.L. - Kinematics of deformation at a thrust fault ramp from syntectonic fibres in pressure shadow. *Geol. Soc. of America Special Paper* **222**, 77 - 88 (1988).
BIOT, M.A. - Theory of folding of stratified viscoelastic media and its implication in tectonics and orogenesis. *Bull. Geol. America Bull.*, **72**: 1595 - 1620 (1961).
BIRCH, F. - Role of fluid pressure in mechanics of overthrust faulting: discussion. *Geol. Soc. America Bull.* **72**, p. 1441 - 1444 (1961).
BLAY, P., COSGROVE, J.W. and SUMMERS, J.M. - An experimental investigation of the development of structures in multilayers under the influence of gravity. *Journ. Geol. Soc. Lond.* **133**, 329 - 342 (1976).
BORGIA, A., FERRARI, L. and PASQUARE, G. - Importance of gravitational spreading in the tectonic and volcanic evolution of Mount Etna. *Nature*, vol. **357**, 231 - 235 (1992).
BOUILLIN, J.P. - *Géologie alpine de al Petite Kabylie dans les régions de Collo et d'El Milia*. Thèse Paris-Toulouse, 511p. 127 fig. (1977).
BOUILLIER, A.M. and QUENARDEL, J.M. - The Caledonides of northern Norway: relation between preferred orientation of quartz lattice, strain and translation of nappes. In: *Thrust and Nappe Tectonics* Spec. Vol Geol.Soc. London **9**, 185 - 195 (1981).
BOYER, S.E. - Style of folding within thrust sheets: examples from the Appalachian and Rocky Mountains of the USA and Canada. *Journ. Struct. Geol.* **8**, 325 - 339 (1986).
BOYER, S.E. and ELLIOTT, D. - Thrust systems. *Bull. Amer. Assoc. Petroleum Geol.* **66**, 1196 - 1230 (1982).
BRESSON, A. - Etude sur les formation anciennes des Jautes et Basses Pyrénées. *Bull. Carte. geol. France* XIV, number 93, 238 - 243 (1903).
BRUN, J.P. - Cumulative strain and boundary effects in the gravity flow of a viscous slab. *Tectonophysics* **41**, T7 - T14 (1977).
BRUN, J.P. - L'origine des dômes gneissiques: modèle et tests. *Bull. Soc. Geol. France* XXV, 219 - 228 (1983).
BRUN, J.P. and MERLE, O. - Strain pattern in models of spreading-gliding nappes. *Tectonics*, **4**, 7, 705 - 719 (1985).
BRUN, J.P. and MERLE, O. - Experiments on folding in spreading-gliding nappes. *Tectonophysics* **145**, 1 - 2, 129 - 139 (1988).
BROCK, W.G. and ENGELDER, T. - Deformation associated with the movement of the Muddy Mountain overthrust in the Buffington Window, South-eastern Nevada.. *Geol. Soc. Am. Bull* **88**, p. 1667 - 1677 (1977).
BUCHER, W.G. - Role of gravity in orogenesis. *Bull. Geol. Soc. Am.* 67, 1295 - 1318 (1956).
BUCHER, W.G. - An experiment of the role of gravity in orogenic folding. *Geo. Rundsch.* **52** (2): 804 - 810 (1962).

BURG, J.P., BALE, P., BRUN, J.P. and GIRARDEAU, J. - Stretching lineation and transport direction in the Ibero-Amorican arc during the Siluro-Devonian collision. *Geodynamica Acta* **1**, 71 - 87 (1987).
BURKHARD, M. - L'Helvétique de la bordure occidentale du massif de l'Aar (évolution tectonique et métamorphique). *Eclog. geol. Helv.* **81**, 63 - 114.
BURKHARD, M. -Aspect of the large-scale Miocene deformation in the most external part of the Swiss Alps (Subalpine Molasse to Jura fold belt). *Eclog. geol. Helv.* **83**, 559 - 583 (1990).
BUTLER, R.W.H. - The terminology of structures in thrust belts. *Journ. Struct. Geol.* **4**, 3, 239 - 245 (1982).
BUTLER, R.W.H. - The restoration of thrust systems and displacement discontinuity around the Mont Blanc Massif, NW external Alpine thrust belt. *Journ. Struct. Geol.* **7**, 569 - 582 (1985).
BUXTORF, A. - Prognosen und Befunden beim hauensteinbasis und Grenchenbergtunnel und die Bedeutung der letztern fur die Geologie des Juragebirges. *Verhandl. Naturforsch. Gesell. Basel* XXVII (1916).
CABY, R., KIENAST, J.R. and SALIOT, P. - Structure, métamorphisme et modèle d'évolution tectonique des Alpes occidentales. *Rev. Geogr. Phys. Geol. Dyn.* **20**, 307 - 322 (1978).
CADELL, H.M. - Experimental researches in mountain building. *Trans. Roy. Soc. Edinburgh* **1**, 337 - 357 (1887).
CALLAWAY, C. - The age of the newer gneissic rocks of the northern highlands. *Quat. Journ. Geol. Soc.* **39**, 355 - 414 (1883).
CHAMBERLIN, R.T. - The Appalachian folds of Central Pennsylvania. *Journ. Geol. Chicago* **18**, 228 - 251 (1910).
CHAMBERLIN, R.T. - Instrumental work on the nature of glacier motion. *Journ. Geol.* **36**, number 1, Chicago (1928).
CHAMBERLIN, R.T. and MILLER, W.Z. - Low-angle faulting. *Journal of Geology* XXVI, **1**, 1-44 (1918).
CHAPPLE, W.H. - Mechanics of thin-skinned fold and thrust belts. *Geol. Soc. Amer. Bull.* 89, 1189 - 1198 (1978).
CHOUKROUNE, P. - Contribution a l'étude des mécanismes de la déformation avec schistosité grâce aux cristallisations syncinematiques dans les zones abritées. *Bull. Soc. Geol. France* XII, number 3 - 4, 257 - 271 (1971).
CHOUKROUNE, P. and SEGURET, M. - Tectonics of the Pyrenees: role of compression and gravity. In: *Gravity and tectonics*, DeJonc, Scholtte ed. John Wiley and Sons, New York, 141 - 156 (1973).
CHOUKROUNE, P. and GAPAIS, D. - Strain pattern in the Aar granite (Central Alps): Orthogneiss developed by bulk in homogeneous flattening. *Jour. Struct. Geol.* **5**, number 4, 001 - 010 (1983).
COBBOLD, P.R. - The restoration of strained cross sections. *Conf. Int. Chevauchement et Deformation*, Toulouse, 15-19 Mai, p. 17 (1984).
COBBOLD, P.R. and SZATMARI, P. - Radial gravitational gliding on passive margins. *Tectonophysics* **188**, 249 - 289 (1991).
COWARD, M.P. - The analysis of flow profiles in a basaltic dyke using strained vesicles. *JH. Geol. Soc. London* **137**, 605 - 615 (1980).
COWARD, M.P. and KIM, J.H. - Strain within thrust sheets. In: *Thrust and Nappe tectonics*, Geological Society Special Publication 9, p. 275 - 292 (1981).
COWARD, M.P. and POTTS, G.J. - Complex strain patterns developed at the frontal and lateral tips to shear zones and thrust zones. *J. Struct. Geol.* **5** (3/4): 383 399 (1983).
CROSBY, G.W. - Radial movement in the western Wyoming Salient of the Cordilleran Overthrust Belt. *Geol. Soc. Aeic Bull.* **80**, 1061 - 1078 (1969).

DAHLEN, F.A. - Noncohesive critical Coulomb wedges: an exact solution. *Journ. Geophys. Res.* **89** (B12), 10125 - 10133 (1984).
DAHLSTROM, C.D.A. - Balanced cross sections. *Can. J. Earth Sci.* **6**, 743 - 747 (1969).
DAHLSTROM, C.D.A. - Structural geology in the eastern margin of the Canadian Rocky Mountains. *Canadian Bull. Petrol. Geol.* **18**, 332 - 402 (1970).
DAL PIAZ, G.B. - L'influenza della gravita nei fenomeni orogenetici. *Atti. R. Acad. Sc. Torino* **77**, (1942).
DAL PIAZ, G.V., HUNZIKER, J.C. and MARTINOTI, G. - La zona Sesia-Lanzo e l'evoluzione tettonico-metamorphica delle Alpi Nordoccidentali interne. *Mem. soc. Geol. Ital.* **11**, 433 - 469 (1972).
DANA, J.D. - Geological results of the earth's contraction in consequence of cooling. *American Journal of Science* III, 176 - 187 (1847).
DAVIS, D.M. and ENGELDER, T. - The role of salt in fold and thrust belts. *Tectonophysics* **119**, 67 - 88 (1985).
DAVIS, D., SUPPE, J. and DAHLEN, F.A. - Mechanics of fold and thrust belts and accretionary wedges: cohesive Coulomb theory. *J. Geophys. Res.* **88** (B2), 1153 - 1172 (1983).
DE BEAUMONT, E. - Recherches sur quelques unes des révolutions de la surface du globe. *Annale Sci. Nat.* XVIII, 5 - 284 (1829).
DE LOYS, F. - Monographie géologique de la Dent du Midi. *Mat. Carte Geol. Suisse* **58** (1928).
DE MARGERIE, E. and HEIM A. - *Les dislocations de l'écorce terrestre*. Zurich, Ed. J. Wurster, 154, (1888).
DENNISON, J.M. and WOODWARD, H.P. - Palinspatic maps of Central Alps. *Bull. Am. Assoc. Petrol. Geol.* **47**, 666 - 80 (1963).
DE PAOR, D.G. - Orthographic analysis of geological structures. Deformation theory. *Journal. Struct. Geol.* 5, number **314**, p. 255 - 277 (1983).
DERAMOND, J. - *Déformation et déplacement des nappes: exemple de la nappe de Gavarnie*. Thèse d'Etat, Toulouse, 409 (1979).
DERAMOND, J. and RAMBACH, J.M. - Mesure de la déformation dans la nappe de Gavarnie (Pyrénées Centrales): interprétation cinématique. *Bull. Soc. Geol. France,* (7), **21** (2): 201 - 211 (1979).
DIETRICH, D. and DURNEY, D.W. - Change of direction of overthrust shear in the Helvetic nappes of western Switzerland. *Journ. Struct. Geol.* **8**, 389 - 398 (1986).
DIETRICH, D. and CASEY, M. - A new tectonic model for the Helvetic nappe. In: *Alpine Tectonics*, Geological Society Special Publication **45**, 47 - 63 (1989).
DIXON, E.E.L. - The Garvarnie overthrust and other problems in Pyrenean geology. *Geol Mag.* **5**, 359 - 374 (1908).
DIXON, J.M. and TIRRUL, R. - Centrifuge modelling of fold-thrust structures in a tripartite stratigraphic succession. *Journ. Struct. Geol.* **13**, 1, 3 - 20 (1991).
DONATH, F.A. and PARKER, R.B. - Folds and folding. *Geol. Soc. Am. Bull.* **75**: 45 - 62 (1964).
DOUGLAS, R.J.W. - Mount head map-area, Alberta. *Geol. Surv. Can. Mem.* 291, - 241p. (1958).
DURNEY, D.W. - Some observations concerning the question of gravity gliding and hinterland compression in the western Helvetic nappes. *International conference on planar and linear fabrics of deformed rocks,* Zurich, Abstract p. 92 (1982).
DURNEY, D.W. and RAMSAY, J.G. - Incremental strains measured by syntectonic crystal growths. In *Gravity and Tectonics* (K.A. de Jong and R. Scholten ed.), John Wiley and Sons ed., New York: 67 - 96 (1973).
EISENSTADT, G. and DE PAOR, D.G. - Alternative model of thrust fault propagation. *Geology* **15**, 630 - 633 (1987).
ELLIOTT, D. - The motion of thrust sheet. *Journal of Geophys. Res.,* 81, 949 - 963 (1976a).

ELLIOTT, D. - The energy balance and deformation mechanisms of thrust sheets. *Phil. Trans. R. Soc. London* **A 283**, 289 - 312 (1976b).
ELLIOTT, D. - Some aspects of the geometry and mechanics of thrust belts. Part I, *8th Annual Seminar Can. Soc. Petrol. Geol. Univ. Calgary* (1977).
ELLIOTT, D. - Mechanics of thin-skinned fold and thrust belts: discussion. *Geol. Soc. Am. Bull.*, part 1, **91**, 185 - 187 (1980).
ELLIOTT, D. - The strength of rocks in thrust sheets. *Eos* **62**, 3997 (1981).
ENDIGNOUX, L. and MUGNIER, J.L. - The use of a forward kinematical model in the construction of a balanced cross section.. *Tectonics* **9**, 1249 - 1262 (1990).
ESCHER, A. ESCHER, J.C. and WATERSON, J. - The reorientation of the Kangamiut dike swarm, West Greenland. *Can. J. Earth Sci.* **12**, 158 - 173 (1975).
ESCHER VON DER LINTH, A., - Verhandl. Schweiz. Naturforsch. Zurich, 54 - 78 (1841).
EVANS, M.E. and DUNNE, W.M. - Strain factorisation and partitioning in the North Mountain thrust sheet, central Appalachian, USA. *Journ. Struct. Geol.* **13**, 1, 21 - 35 (1991).
FALLOT, P. - Observation sur la tectonique de la zone subbetique dans la province de Murcia. *Bull. Soc. Geol. France* XIV, 11 - 28 (1944).
FALLOT, P. - Maurice Lugeon (1870 - 1953). *Bull. Soc. Geol. France* (6), 4 303 - 340 (1954).
FAVRE, A. - Expériences sur les effets des refoulements ou écrasement latéraux en géologie. *C. R. Acad. Sci. Paris* XXXVI, 1092 - 1095 (1878).
FLETCHER, P. and GAY, N.C. - Analysis of gravity sliding and orogenic translation: discussion. *Geol. Soc. Am. Bull.***82**, p. 2677 - 2682 (1971).
FLINN, D., - On folding during three dimensional progressive deformation. *Quart. J. Geol. Soc. Lond.* **118**, 385 - 434 (1962).
FORRISTALL, G.Z. - Stress distributions and overthrust faulting. *Geol. Soc. Am. Bull* **83**, 3073 - 3081 (1972).
FOUCAULT, A. and RAOULT, J.F. - *Dictionnaire de géologie*. Masson, 331p. (1980).
FOURMARIER, P. L'évaluation de l'importance des phénomènes de charriages en Belgique et dans les régions voisines. *Extrait du Comptes-Rendus du XIIIe Congres géologique international* (1922). Liège, imprimerie Vaillant-Carmanne (1923).
GAMOND, J.F. - Direction de déplacement et lineation: cas de la couverture sédimentaire dauphinoise orientale. *Bull. Soc. Geol. France* XXII, 429 - 436 (1980).
GEIKIE, A. - The crystalline rocks of the Scottish highlands. *Nature* XXXi, 29 - 31 (1884).
GEIKIE, A . - *Textbook of Geology*. 4 volumes, New York (third edition) (1893).
GEISER, P.A. - Mechanisms of thrust propagation: some examples and implications for the analysis of overthrust terranes. *Journ. Struct. Geol.***10**, 8, 829 - 845 (1988).
GESE, B. - Distinction d'un type de nappes à enracinement frontal: les refoulements. *C. R. Sommaire des séances de la société géologique de France*, number 2, p. 38 (1962).
GIDON, M. - *Les structures tectoniques*. BRGM, Collection Manuels and Méthodes, 206p. (1987).
GIGNOUX, M. - Méditation sur la théorie de la tectonique d'écoulement par gravité. *Trav. Lab. Geol.* **27**, 1- 34 (1948).
GILBERT, E. and MERLE, O. - Extrusion and radial spreading beyond a closing channel. *Journ. Struct. Geol* **9**, 4, 481 - 490 (1987).
GILLCRIST, R., COWARD, M. and MUGNIER, J.L. - Structural inversion and its control: examples from the Alpine foreland and the French Alps. *Geodyn. Acta* **I** (1): 5 - 34 (1987).
GILLET, Ph., CHOUKROUNE, P. BALLEVRE, M. and DAVY, PH. - Thickening history of the Western Alps. *Earth Planet Sci. Lett.* **78**, 44 - 52 (1986).
GOGUEL, J. - Introduction à l'étude mécanique des déformations de l'écorce terrestres (2e ed.) *Mem. Expl. Carte Geol. Fr.*, 530p. (1948).
GOGUEL, J. - *Traite de tectonique*. Editeur Masson et Cie, 383p. (1952).

GOGUEL, J. - Le rôle de l'eau et de la chaleur dans les phénomènes tectoniques. *Rev. Geogr. Phys. Geol. Dynam.* **23**, 177 - 193 (1969).
GOHAU, G. - *Histoire de la géologie*. Edition La Découverte (1987).
GOSSELET, J. - Sur la structure générale du bassin houiller franco-belge. *Bull. Soc. Gel. France* VIII, 505 - 512 (1879).
GOULD, S.J. - *Le sourire du flamand rose.*, Edition française: Seuil (1988).
GOURLAY, P. - La déformation du socle et des couvertures Delphino-Helvetiques dans la région du Mont-Blanc (Alpes Occidentales). *Bull. Soc. Geol. France* **1**: 159 - 169 (1986).
GRAHAM, R., HOSSACK, J., DERAMOND, J. and SOULAS, J.C. - Géométrie des surfaces de chevauchements. *Bull. Soc. Geol. France* III, number 1, 169 - 181 (1987).
GRANT, N.T. - Post-emplacement extension within a thrust sheet from the Central Pyrenees. *Journ. Geol. Soc. London* **149**, 775 - 792 (1992).
GRATIER, J.P. - L'équilibrage des coupes géologiques. *Mémoires et documents du Centre Armoricain d'Etude Structurale des Socles* **20**, 157p. (1988).
GRATIER, J.P., MENARD, G. and ARPIN, R. - Strain-displacement and restoration of the Cain Subalpines of the Western Alps. In: *Alpine Tectonics*, Geological Society Special Publication 45, 65 - 81 (1989).
GRATIER, J.P., GUILLIER, B., DELORME, A. and ODONNE, F. - Restoration and balance of folded and faulted surface by best-fitting of finite elements: principle and application. *Journ. Struct. Geol.* **13**, 1, 111 - 115 (1991).
GRAY, D.R. and WILLMAN, C.E. - Thrust related strain gradients and thrusting mechanism in chevron-folded sequence, south-eastern Australia. *Journ. Struct. Geol.* **13**, 6, 691 - 710 (1991).
GRETENER, P.E. - Pore pressure, discontinuities, isostasy and overthrust. In: *Thrust and nappe tectonics*, Geological Society Special Publication **9**, 33 - 39 (1981).
GRIGGS, D.T. - A theory of mountain building. *American Journal of Science* CCXXXVII, p. 611 (1939).
GROSHONG, R.H. and USKANSKY, S.I. - Using a micro-computer for interactive section construction and balancing (abs). *Bull. Amer. Assoc. Petrol. Geol.* **70**, 59 (1986).
GUCWA, P.R. and KELHE, R.O. - Bearpaw Mountains rockslide, Montana, USA. In: *Rockslides and Avalanches,* Voight, B., ed. Natural Phenomena: Amsterdam, Elsevier, 393 - 421 (1978).
GUILLIER, B. - *Modélisation analogique des interactions compression gravite.* Diplôme d'Etude Approfondie, Rennes, p. 36 (1988).
GUILLIER, B. - *Dépliage automatique de strates plissées et faillées: application a l'équilibrage de structures naturelles.* Thèse d'Université, Institute de Recherche Interdisciplinaire de Géologie et de Mécanique (Grenoble), p. 160 (1991).
GUTERMAN, V.G. - Model studies of gravitational gliding tectonics. *Tectonophysics* **65**, p. 111 - 126 (1980).
GWINN, V.E. - Kinematic patterns and estimates of lateral shortening, Valley and Ridge and Great Valley provinces, in: *Studies of Appalachians Geologie: Central and Southern Central Appalachians, south-central Pennsylvania.* In Fisher, G.W. et al. 5eds, Wiley, New York, 127-146 (1970).
HAARMANN, E. - *Die oszillations theorie; eine erklarung der krustenbewegung von erde und mond.* F. Enke, Stuggart (1930).
HAFNER, W. - Stress distribution and faulting. *Bull. Soc. Geol. Am.* **62**, 373 - 398 (1951).
HAMBREY, M.J. - Foliation, minor folds and strain in glacier age. *Tectonophysics* **39**, 397 - 416 (1977).
HARRIS, L.B., BURG, J.P. and SAUNIAC, S. - Strain distribution within the Parpaillan nappe (Montagne Noire, France) and structure of its basal thrust zone: implications for events associated with nappe emplacement. *Journ. Struct. Geol.* **5**, 3/4, 431 - 440 (1983).

HAYES, C.W. - The overthrust faults of the southern Appalachians. *Geol. Soc. Am. Bull.* **2**, 141 - 154 (1891).
HEARD, H.C. - Effect of changes in strain rate in the experimental deformation of the Yule marble. *Journal Geol.* **71**, 162 - 195 (1963).
HEARD, H.C. and RUBEY, W.W. - Tectonic implication of gypsum dehydration. *Geol. Soc. Am. Bull* **77**, 741 - 760 (1966).
HEIM, A. - *Untersuchungen u ber den Mechanismus der Gebirgsbildung.* Basel (1878).
HEIM, A. - Geologie der Hochalpen zwischen Reuss und Rhein. *Beitrage Geol. Karte Schweiz* (1891).
HEIM, A. - *Geologie der Schwiez.* (II/4) Tauchnitz, Leipzig (1919 - 1922).
HOBBS, W.H. - Notes on the English equivalent of Schuppenstruktur. *Journ. Geol.* **1**, p. 206 (1894).
HOLMES, A. - The thermal history of the Earth. *Journ. of the Washington Academy of Sciences* XXIII, p. 169 (1933).
HOMEWOOD, P., ALLEN, P.A. and WILLIAMS, G.D. - Dynamics of the Molasse Basin of western Switzerland. *Spec. Publs int. Ass. Sediment.* **8**, 199 - 217 (1986).
HOMEWOOD, P. and LATELTIN, O. - Classic Swiss clastics (flysch and molasse); the Alpine connection. *Geodinamica Acta* (Paris) **2**, 1 - 11 (1988).
HOSSACK, J.R. - The use of balanced cross sections in the calculation of orogenic translation: a review. *J. Geol. Soc. Lond.* **136**, 705 - 711 (1979).
HSU, K.J. - Role of cohesive strength in the mechanics of overthrust faulting and of landsliding. *Geol. Soc. Am. Bull* **80**, 927 - 952 (1969).
HSU, K.J. - Exhumation of high-pressure metamorphic rocks. *Geology* **19**., 107 - 110 (1991).
HSU, T.C. - Velocity field and strain rates in plastic deformation. *J. Strain Analysis* **2**, 196 - 206 (1967).
HUBBERT, M.K. - Mechanical basis for certain familiar geological structures. *Geol. Soc. Am. Bull.* **62**, 335 - 372 (1951).
HUBBERT, M.K. and RUBEY, W.W. - Role of fluid pressure in mechanics of overthrust faulting, I. mechanics of fluid-filled porous solids and its application to overthrust faulting. *Geol. Soc. Am. Bull* **70**, 115 - 166 (1959).
HUDLESTON, P.J. - Recumbent folding in the base of the Barnes Ice Cap, Baffin Islands, Northwest Territories Canada. *Geol. Soc. Am. Bull.* 87, 1684 - 1692 (1976).
HUDLESTON, P.J. - Similar folds, recumbent folds and gravity tectonic in ice and rocks *Journ Geol.* **85**, 113 - 122 (1977).
HUDLESTON, P.J. and HOOKE, R.F. - Cumulative deformation in the Barnes ice cap and implication of the development of foliation. *Tectonophysics* 66., 127 - 146 (1980).
HUNT, C.W. - Planimetric equation. *Journ. Alberta Pet. Geol.* **56**, 259 - 264 (1957).
HUPPERT, H. - Flow and instability of a viscous current down a slope. *Nature* 300 (589) (1982).
JAMISON, W.R. - Geometric analysis of fold development in overthrust terranes. *Journ. Struct. Geol.*9, 2, 207 - 219 (1987).
JEANBOURQUIN, P. and GOY-EGGENBERGER, D. - Melange suprahelvetiques: sedimentation et tectonique au front de la nappe de Morcles (Vaud, Suisse). *Geologie Alpine* **67**, 43 - 62 (1991).
JOHNSON, M.R.W. - The erosion factor in the emplacement of the Keystone Thrust Sheet (South East Nevada) across a land surface. *Geol. Mag.* **118**, 501 - 507 (1981).
JONES, P.B. - Quantitative geometry of a thrust and fold belt structures. *Amer. Assoc. Petrol. Geol.*, Tulsa, Oklahoma, USA (1987).
JONES, P.B. and LINNSER, H. - Computer synthesis of balanced structural cross-sections by forward modelling (abs.) *Bull. Amer. Assoc. Petrol. Geol.* **70**, 605 (1986).

KELHE, R.O. - Analysis of gravity sliding and orogenic translation. *Bull. Geol. Soc,. Amer.* 8?, 1641 - 1664 (1970).
KELHE, R.O. - Analysis of gravity sliding and orogenic translation: reply. *Bull. Geol. Soc,. Amer.* **82**, 2683 - 2684 (1971).
KERCKHOVE, Cl. - La zone du flysch dans les nappes de l'Embrunais - Ubaye. *Geol. Apline* **45**, 5 - 204 (1969).
KIEFFER, J.D. and DENNISON, J.M. - Palinspatic map of Devonian strata of Alberta and Northwest Georgia. *Bull. Am. Assoc. Petrol. Geol.* 56, 161 - 166 (1972).
KLIGFIELD, R., GEISER, P. and GEISER, J. - Construction of geological cross-sections using microcomputer system. *Geobyte* **1**, 60 -66 (1986).
KLIGFIELD., R., CARMIGNANI, L. and OWENS, W.H. - Strain analysis of a northern Apennine shear zone using deformed marble breccias. *Journ. Struct. Geol.***3**, number 4, 421 - 436 (1981).
KNIPE, R.J. - Footwall geometry and the rheology of thrust sheet. *Journ. Struct. Geol* 7, 1, 1 - 10 (1985).
KUHN, T.S. - *La structure des révolutions scientifiques.* Edition Française: Flamarion (1983).
LAPWORTH, C. - The Moffat series. *Qat. Journ. Geol. Soc.* **30**, 240 - 310 (1878).
LAUBSCHER, H.P. - Die Fernschubhypothese der Jurafaltung. *Eclog. Geol. Helv.* **54** (1), 221 - 282 (1961).
LAUBSCHER, H.P. - Viscous component in Jura folding. *Tectonophysics* 27, 239 - 254 (1975).
LAW, R.D., KNIPE, R.J. and DAYAN, H. - Strain path partitioning within thrust sheet : microstructural and petrofabric evidence from the Moine Thrust at Loch Eriboll, north-w ,r Scotland. *Journ. Struct. Geol.* **6**, 477 - 497 (1984).
LEHNER, F.K. - Comments on 'noncohesive critical Coulomb wedges: an exact solution' by F.A. Dalhen, *Journal of Geophys. Res.,***91**, (B1), 793 - 796 (1986).
LEMOINE, M. - About gravity gliding tectonics in the Western Alps. in: *Gravity and tectonics.* Edited by K.A. De Jong and R. Scholten, John Wiley, New York, p. 201 - 216 (1973).
LOGAN, W.E. - Consideration relating to the Quebec Group and the Upper Copper-bearing rocks of Lake Superior. *American Journal of Science* XXXIII, second series, 320 - 327 (1862).
LOOSVELD, R. and SCHREURS, G. - Discovery of thrust klippen, north-west of Mary Kathleen, Mt. Isa Inlier, Australia. *Australian Journal of Earth Science* **34**, 387 - 402 (1987).
LUGEON, M. - Les dislocations des Bauges (Savoie*). Bull. Serv. Carte geol. France* XI, number 77, 359 - 470 (1901).
LUGEON, M. - Les nappes de recouvrement de la Tatra et l'origine des klippes des Carpathes. *Bull. Soc. vaud. Sc. nat.* XXXIV (1903).
LUGEON, M. - Une nouvelle hypothèse tectonique: la divertiulation. *Bull. Soc. Vaud. Sc. Nat.* **61**, number 260 (1943).
LUGEON, M. and GAGNEBIN, E. - Observations et vues nouvelles sur la géologie des Prealpes Romandes. *Bull. Lan. Geol. Univ. Lausanne* **72**, (1941).
LIU HUIQI, K.R., McCLAY, K.R. and POWELLL, D. - Physical models of thrust wedges. In *Thrust tectonics*, edited by K.R. McClay, Chapman and Hall, 71 - 81 (1992).
LUNDIN, E.R. - Thrusting of the Claron formation, the Bryce canyon region. *Bull. Soc. Geol. Amer.* **101**, 1038 - 1050 (1989).
McCONNEL, R.G. - Report on the geological structure of a portion of the Rocky Mountains. *Geol. Surv. Canada,* Annual Report for 1886, part D (1886).
MAILLET, R. and PAVANS DE CECCATY, R. - Le physicien devant la tectonique. *Congres mondial du Pétrole*, Paris (1937).
MALAVIEILLE, J. - Modélisation expérimentale des chevauchements imbriques: application aux chaînes de montagnes. *Bull. Soc. Geol. France* **26**, 129 - 138 (1984).
MALAVIEILLE, J. and RITZ, J.F. - Mylonitic deformation of evaporites in decollements: examples from the southern Alps, France. *Journ. Struct. Geol.* **11**, 5, 583 - 590 (1989).

MALVERN, L.E. - *Introduction to the mechanics of a continuous medium*. Prentice-Hall, inc. Englewood Cliffs, New Jersey, 713p. (1969).
MANDL, G. and SHIPPAM, G.K. - Mechanical model of thrust sheet gliding and imbrication. In *Thrust and nappe tectonics*, Geological Society Special Publ. **9**, 79 - 98 (1981).
MANDL, G. - *Mechanics of tectonic faulting, Models and basic concepts*. Series Editor: J.J. Zwart, Development in Structural Geology 1, Elsevier Amsterdam, 407p. (1988).
MARCOUX, J, BRUN, J.P., BURG, J.P. and RICOU, L.E. - Shear structures in anhydrite at the base of thrust sheets (Antalya, southern Turkey). *Journ. Struct. Geol* **9**, 555 - 561 (1987).
MARSHACK, S. and ENGELDER, T. - Development of cleavage in limestones of fold-thrust belt in eastern New York. *Journ. Struct. Geol.* **7**, 3/4, 345 - 359 (1985).
MASSON,H. - Sur l'orogine de la cargneule par fracturation hydraulique. *Eclog. Geol. Helv.* **65**, number 1, 27 - 41 (1972).
MASSON,H. - Un siècle de géologie des prealpes: de la découverte des nappes à la recherche de leur dynamique. *Eclog. Geol. Helv.* **69**, 527 - 575 (1976).
MASSON,H. - La géologie en Suisse de 1882 a 1932. *Eclog. geol. Helv.* 69/2, 47 - 64 (1983).
MATTAUER, M. - Etude géologique de l'Ouarsenis oriental (Algérie). *Publication du Service de al carte géologique de l'Algérie* 17, 534p. (1958).
MATTE, Ph. - La structure de la virgation hercynienne de Galice (Espagne). *Trav. Lab. Geol. Grenoble* **44**, 128p. (1968).
MEDWEDEFF, D.A. and SUPPE, J. - Kinematics, timing and rates of folding and faulting from syntectonic sediments geometry. *EOS* **67**, 1223 (1986).
MENARD, G. - Méthodologie générale de construction des coupes équilibrées. *Mémoires et Documents du Centre Armoricain d'Etude Structurales des Socles* **20**, 5 - 25 (1988).
MERCIER, J. and VERGELY, P. - *Tectonique*. Collection Geosciences, dunod, 214p. (1992).
MERLE, O. - *Cinematique et déformation de la nappe du Parpaillon (flysch a Helminthoides, Alpes Occidentales)*, PhD thesis 147p. (1982)
MERLE, O. - Déplacement et déformation des nappes superficielles. *Rev. Geol. Dyun. Geogr. Phys.* **25**, 3 - 17 (1984).
MERLE, O. -Patterns of stretch trajectories and strain rate within spreading-gliding nappes. *Tectonophysics* **124**, 3/4, 211 - 222 (1986).
MERLE, O. -Strain pattern within spreading nappes. *Tectonophysics* **165**, 57 - 71 (1989).
MERLE, O. - Cinématique des nappes superficielles et profondes dans une chaîne de collision. *Mémoires et Documents du Centre Armoricain d'Etudes Structurales des Socles* **37**, 280p., 165 fig., 3 tabl. (1990).
MERLE, O. and BRUN, J.P. - The curved translation path of the Parpaillon nappe (French Alps). *Journ. Struct. Geol.***6**,6, 711 - 719 (1984).
MERLE, O. and GUILLIER, B. - The building of the Swiss Central Alps: an experimental approach. *Tectonophysics* **165**, 41 - 56 (1989).
MERLE, O. and VENDEVILLE, B. - Experimental modelling of thin-skinned shortening around magmatic intrusions. *Bulletin of Volcanology* **57**: 33 - 43 (1995).
MERLE, O. and BALLEVRE, M. - Late Cretaceous-early Tertiary detachment fault in the Western Alps. *C. R. Acad. Sci. Paris* 315, II, 1769 - 1776 (1992).
MERLE, O., NICKELSEN, R.P., DAVIS, G.H. and GOURLAY, P. - Relation of thin-skinned thrusting of Colorado Plateau strata in south-western Utah to Cenozoic magmatism. *Bull. Geol. Soc. Amer.* **105**, 387 - 398 (1993).
MERLE, O. and ABIDI, M.N. - Approche expérimentale du fonctionnement des rampes émergentes. *Bull. Soc. Geol. France* **166**, 5, 439 - 450. (1995).
MEYERHOFF, A.A., TANER, I., MARTIN, A.E., AGOCS, W.B. and MEYERHOFF, H.A. - Surge tectonics: a new hypothesis of Earth dynamics. In Chatterjee, S. and N. Norton III, eds. *New Concepts in Global Tectonics*, Texas Tech University Press, Lubbock, 309 - 409 (1992).

MICHARD, A., LE MER, O., GOFE, B. and MOTIGNY, R. - Mechanism of the Oman mountains obduction onto the Arabian continental margin, reviewed. *Bull. Soc. Geol. France* **8**, V, 241 - 252 (1989).

MILTON, M.J. and WILLIAMS, G.D - The strain profile above a major thrust fault, Finnmark, Norway. In: *Thrust and Nappe Tectonics* Geological Society Special Publication 9, 235 - 239 (1981).

MITRA, S. - Duplex structure and imbricate thrust systems: geometry, structural position and hydrocarbon potential. *AAPG Bull.* **70/9**, 1087 - 1112 (1986).

MITRA, G. and ELLIOTT, D. - Deformation of basement in the Blue Ridge and the development of the South Mountain cleavage. In: *The Caledonides in the USA*, D.R. Wones (Editor), Va. Polytech. Inst. State Univ., Mem., **2**: 307 - 311 (1980).

MITRA, G. and BOYER, S.E. - Energy balance and deformation mechanisms of duplexes. . *Journ. Struct. Geol* **8**, 291 - 304 (1986).

MORET, L. - Les idées nouvelles sur l'origine des chaînes de montagnes. *Trav. Lab. Geol. Grenoble* XXVIII, 1 - 56 (1950).

MORLEY, C.K. - A classification of thrust fronts. *Amer. Assoc. Petrol. Geol. Bull.* **70**, 1, 12 - 25 (1986).

MORSE, J. - Deformation in ramp regions of overthrust faults: experiments with small scale rock models. Guidebook *Wyo. Geol. Assoc. field conf.* 29th, 457 - 470 (1977).

MUGNIER, J.L. - *Déplacements et déformations dans l'avant pays d'une chaîne de collision. Méthodes d'études et modélisation. Exemple du Jura.* Thèse Docteur Ingénieur, Grenoble, 163p. (1984).

MUGNIER, J.L., MASCLE, G. and FAUCHER, T. - La structure des Swaliks de l'Ouest Népal: un prisme d'accrétions intracontinental. *Bull. Soc. geol. France*, **163**, 585 - 595 (1992).

MUGNIER, J.L. and VIALON, P. - Deformation and displacement of the Jura cover on its basement. . *Journ. Struct. Geol* **8**, 3/4, 373 - 388 (1986).

MULUGETA, G. - Modelling the geometry of Coulomb thrust wedges. *Journ. Struct. Geol* **10**, number 8, 847 - 859 (1988).

NAEF, H., DIEBOLD, P. and SCHLANKE, S. - Sedimentation und tektonik im Tertiar der Nordschweiz. *Nagra Techn. Ber.* 85 - 14, Nagra, Baden (1985).

NICKELSEN, R.P. - Attributes of rock cleavage in some mudstones and limestones of the Valley and Ridge Province, Pennsylvania. *Proceedings of the Pennsylvania Academy of Science* **46**, 107 - 112 (1972).

NICKELSEN, R.P. - Cleavage duplexes in the Marcellus shale of the Appalachian foreland. *Journ. Struct. Geol* **8**, 3/4, 361 - 371 (1986).

NICOLAS, A. - *Principe de tectonique.* Masson, 196p. (1984).

NICOLAS, A. - *Structures of ophiolites and dynamics of oceanic lithosphere*, Kluwer Acad. Publ., Dordrecht 368p. (1989).

NYE, J.P. - The mechanics of glacier flow. *Journ. Glaciology* **2**, 82 - 93 (1952).

ODONNE, F. and VIALON, P. - Analogue models of folds above a wrench fault. *Tectonophysics* **99**, 31 - 46 (1983).

PEACH, B.N. and HORNE, J. - Report on the geology of the North-West of Sutherland. *Nature* XXXI, 31 - 35 (1884).

PHILIPPOT, P. - Déformation et eclogitisation progressives d'une croûte océanique subductee: le Monviso, Alpes occidentales. Contraintes cinématique durant la collision Alpine. *docu. Tra Centre Geol. Geophys. Montpellier* **19**, 270p. (1988).

PLANCHEREL, T. - Aspect de la déformation en grand dans les prealpes médianes plastiques entre Rhône et Aar. *Eclogae Geol. Helv.* **72**, 1, 145 - 214 (1979).

PLATT, J.P. - Dynamics of orogenic wedges and the uplift of high-pressure metamorphic rocks. *Geol. Soc. Amer. Bull.* **97**, 1037 - 1053 (1986).

PLATT, L.B. - Fluid pressure in thrust faulting: a corollary. *Amer. Jour. Sci.* 260, 107 - 114 (1961).
POTTS, G.J. - Finite strain within recumbent folds of Kishorn nappe, Northwest Scotland. *Tectonophysics* **88**, 313 - 319 (1982).
PRICE, N.J. and JOHNSON, M.R.W. - A mechanical analysis of the Keystone-Muddy mountain thrust sheet in south-east Nevada, *Tectonophysics* **84**, 131 - 150 (1982).
PRICE, N.J. and COSGROVE, J.W. - *Analysis of geological structures.* Cambridge University Press, Cambridge 502p. (1990).
PRICE, R.A. and MOUNTJOY, E.W. - Geologic structure of the Canadian Rockies Mountains between Bow and Athabasca rivers: a progress report. *Geol. Assoc. Can. Sp. Pap.* **6**, 7 - 24 (1970).
PRICE, R.A. - Large scale gravitational flow of supracrustal rocks, southern Canadian Rockies. In: *Gravity and tectonics* (De Jong, Scholten ed.) John Wiley and Sons, New York, 491 - 502, (1973).
QUINQUIS, H., AUDRUN, C., BRUN, J.P. and COBBOLD, P.R. - Intense progressive shear in Ile de Groix: blueschists and compatibility with subduction or obduction. *Nature* **273** (5657): 43 - 45 (1978).
RALEIGH, C.B. and GRIGGS, D.T. - Effects of the toe in the mechanics of overthrust faulting. *Geol. Soc. Am. Bull.* **74**, 819 - 830 (1963).
RAMBACH, J.M. and DERAMOND, J. - Constant thickness overthrust on a visco-plastic sole. *Tectonophysics* **60**, T7 - T17 (1979).
RAMBERG, H. - Contact strain and folding instability of a multilayered body under compression. *Geol. Rundsch.* 51, 405 - 439 (1961).
RAMBERG, H. - Note on model studies of folding of moraines in Piedmont glaciers. *J. Glaciol* **5**, 207 - 218 (1964).
RAMBERG, H. - Particle paths, displacement and progressive strain applicable to rocks. *Tectonophysics* **28**, 1 - 37 (1975).
RAMBERG, H. - Superposition of homogeneous strain and progressive deformation in rocks. *Bull. Geol. Inst. Univ. Uppsala* **6**, 35 - 67 (1975).
RAMBERG, H. - Some remarks on the mechanism of nappe movement. *Geol. Foremingen Stockholm Formandlingar* 110 - 117 (1977).
RAMBERG, H. - *Gravity, deformation and the Earth's crust.* Academic Press, London, 452p. (1981).
RAMSAY, J.G. - *Folding and fracturing of rocks* McGraw Hill, New York, 568p (1967).
RAMSAY, J.G. - The measurement of strain and displacement in orogenic belts. In *Time and Place in Orogeny.* Geol. Soc. London, 43 - 79 (1969).
RAMSAY, J.G. - Tectonics of the Helvetic nappes. In *Thrust and Nappe Tectonics,* Geological Society Special Publication **9**, 293 - 303 (1981).
RAMSAY, J.G., CASEY, M. and KLIGFIELD, R. - Role of shear in development of the Helvetic fold-thrust belt of Switzerland. *Geology* **11**, 439 - 442 (1983).
RATSCHBACHER, L., WENK, H.R. and SINTUBIN, M. - Calcite textures: example from nappes with strain-path partitioning. *Journ. Struct. Geol* **13**, number 4, 369 - 384 (1991).
RATTEY, P.R. and SANDERSON, D.J. - Patterns of folding within nappes and thrust sheet: example from the Variscan of Southwest England. *Tectonophysics* **88** (3/4): 247 - 267 (1982).
READE, T.M. - The mechanics of overthrusts. *Geol. Mag.* **5**, 518 (1908).
REEVES, F. - Shallow folding and faulting around the Bearpaw Mountains. *American Journal of Sciences* **10** 187 - 200 (1925).
REKS, I.J. and GRAY, D.R. - Strain patterns and shortening in a folded thrust sheet: an example from the southern Appalachians. *Tectonophysics* **93**, 99 - 128 (1983).
REYER, E. - *Theoritische Geologie.* Stuttgart 868p. (1888).

RHODES, S. and GAYER, R.A. - Non-cylindrical folds, linear structures in the X direction and mylonite developed during translation of the Caledonian Kalak nappes complex of Finmark. *Geol. Mag.* **114** (5), 329 - 341 (1977).

RICH, J.L. - Mechanics of low-angle overthrust faulting as illustrated by Cumberland thrust block, Virginia, Kentucky, and Tennessee. *American Association of Petroleum Geologists Bulletin* **18**, 1584 - 1587 (1934).

RIDLEY, J. and CASEY, M. - Numerical modelling of folding in rotational strain histories: strain regime expected in thrust belts and shear zones. *Geology* **17**, 875 - 878 (1989).

RIGASSI, D. - Genèse tectonique du Jura: une nouvelle hypothèse. *Paleolab.* News 2, Terreaux du Temple, Genève.

ROD, E. - Mechanics of thin-skinned fold and thrust belts. Discussion. *Bull Geol. Soc. Am.* **91**, 188p. (1980).

ROEDER, D., GILBERT Jr., O.E. and WITHERSPOON, W.D. - Evolution and macroscopic structure of Valley and Ridge thrust belt, Tennessee and Virginia. *Stud. in Geol.* **2**, 25p. dep. of Geol. Sci., Univ. of Tenn. Chattanooga (1978).

ROGERS, H.D. and ROGERS, W.B. - On the physical structure of the Appalachian chain, as exemplifying the laws which have regulated the elevation of great mountain chains generally. Rep. 1st to 3rd Meetings Ass. Amer. Geol. and Nat., Boston; abstract in *Rep. Brit. Ass.*, Manchester, 1842, p. 40 (1843).

ROUBAULT, M. - La genese des montagnes. *PUF*, 243p. (1949).

RUBEY, W.W. and HUBBERT, M.K. - Role of fluid pressure in mechanics of overthrust faulting. *Geol. Soc. Am. Bull.* **70**, 167 - 205 (1959).

RUTTER, E.H. - The kinetics of rock deformation by pressure solution. *Phil. Trans. R. Soc. London* A283, 203 - 219 (1976).

SANDERSON, D.J. - The transition from upright to recumbent folding in the Variscan fold belt of south-west England: a model based on the kinematics of simple shear. *Journ. Struct. Geol* **1**, 171 - 180 (1979).

SANDERSON, D.J. - Models of strain variation in nappes and thrust sheets: a review. *Tectonophysics* **88**, 201 - 233 (1982).

SANDERSON, D.J., ANDREWS, J.R., PHILIPPS, W.E.A. and HUTTON, D.H.W. - Deformation studies in the Irish Caledonides. *J. Geol. Soc. London* **137**, 289 - 302 (1980).

SCHARDT, H. - Sur l'origine des Prealpes romands. *Arch. Sci. phys. nat* (3), **30**, 570 - 583 (1893).

SCHARDT, H. - Les régions exotiques du versant Nord des Alpes Suisse. *Bull. Soc. Vaud. Sci. Nat.* **34**, 114 - 219 (1898).

SCHMID, S.M. - The Glarus overthrust: field evidence and mechanical model. *Eclog. Geol. Helv.* **68**, 247 - 280 (1975).

SCHNEEGANS, D. - La géologie des nappes de l'Embrumais Ubaye entre la Durance et l'Ubaye. *Mem. carte Geol.*, France (1938).

SCHOLTEN, R. - Gravitational mechanism in the Northern Rocky Mountains of the United States. In *Gravity and Tectonics*, Edited by K.J. De Jong and R. Scholten, John Wiley New York, 473 - 489 (1973).

SCHORBORN, G. - Alpine tectonics and kinematic models of the Central Southern Alps. *Memorie de Scienze Geologische* XLIV, 229 - 393 (1992).

SEGURET, M. - Etude tectonique des nappes et séries décollées de la partie centrale du versant sud des Pyrénées; caractère synsedimentaire, rôle de al compression et de la gravite. Publ. *USTELA, Série Géologie Structurale* **2**, 155p. (1972).

SENO, S. - Finite strain and deformation within the Bianconnais Castelvecchio-Cerisola nappe of the Ligurian Alps, Italy. *Journ. Struct. Geol* **14**, 7, 825 - 838 (1992).

SHAW, J.H. and SUPPE, J.(1994)- Active faulting and growth folding in the eastern Santa Barbara Channel, California. Geol.Soc. Amer. Bull. 106, 607-626.
SIDDANS, A.W.B. - Arcuate fold and thrust patterns in the subalpine chains of Southeast France. *Journ. Struct. Geol* **1**, number 2, 117 - 126 (1979).
SIDDANS, A.W.B. - Finite strain patterns in some Alpine nappes. *Journ. Struct. Geol* **5**, 441 - 445 (1983).
SIDDANS, A.W.B. - Thrust tectonics. A mechanistic view from the West and Central Alps. *Tectonophysics* **104**, 257 - 281 (1984).
SKJERNAA, L. - Rotation and deformation of randomly oriented planar and linear structures in progressive simple shear. *Journ. Struct. Geol* **1**(1), 5 - 18 (1980).
SMITH, A.G. - Subduction and coeval thrust belts, with particular reference to North America. In: *Thrust and nappe tectonics*, Geological Society Special Publication **9**, 111 - 124 (1981).
SMOLUKOWSKY, M.S. - Some remarks on the mechanics of overthrust. *Geol. Mag.* **6**, 203 - 205 (1909).
SOLLAS, W.J. - Recumbent folds produced as a result of flow. *Quart. Journ. Geol. Soc. London* **62**, 716 - 721 (1906).
SPENCER, S. - *The nature of the north Pennine front: French Alps.* PhD thesis, University of London. (1989).
STAUB, R. - Der bau der Alpen: versuch einer synthese. *Beitr. Geol. karte schweiz*, NF 52, (1924).
STECK, A. - Le massif du Simplon: reflexions sur la cinematique des nappes de gneiss. *Schweiz. Mineral. Petrogr. Mitt.* **67**, 27 - 45 (1987).
STOCKMAL. G.S. - Modelling of large-scale accretionary wedge formation. *Journ. Geophys. Res.* **88**, 8271 - 8287 (1983).
SUESS, E. - *Die Entstehung der Alpen*. Vienna (1875).
SUESS, E. - *Das antlitz der Erde*. 3 volumes, Vienna (1883 - 1909).
SUPPE, J. - Geometry and kinematics of fault-bend folding. *American Journal of Science* **283**, 684 - 721 (1983).
SUPPE, J. - *Principles of Structural Geology.* Englewood Cliffs, New Jersey, Prentice-hall, 537p. (1985).
SUPPE, J., CHOU, G.T. and HOOK, S.C. - Rates of folding and faulting determined from growth strata. In *Thrust Tectonics* ed. McClay K.R., Chapman and Hall, 105 - 121 (1992).
TABOADA, A., BOUSQUET, J.C. and PHILIP, H. - Coseismic models of folds above blind thrusts in the Betic Cordilleras (Spain) and evaluation of seismic hazard. *Tectonophysics* **220**, 223 - 241 (1993).
TABOADA, A., RITZ, J.F. and MALAVIEILLE, J. - Effect of ramp geometry on deformation in a ductile decollement level. *Journ. Struct. Geol.* **12**, number 3, 297 - 302 (1990).
TALBOT, C.J. - Fold trains in a glacier of salt in Southern Iran. *Journ. Struct. Geol.* 1, number 1, 5 - 18 (1979).
TANNER, P.W.G. - Morphology and geometry of duplexes formed during flexural-slip folding. *Journ. Struct. Geol* **14**, 10, 1173 - 119 (1992).
TARDY, M., DEVILLE, E., FUDRAL, S., GUELLEC, S., MENARD, F., THOUVENOT, F. and VIALON, P. - Interprétation des données du profil de sismique réflexion profonde ECORS-CROP Alpes entre le front pennique et la ligne du Canavese (Alpes occidentales). *Mem. Soc. Geol. France* **156**, 217 - 226 (1990).
TERMIER, P. - *A la gloire de la terre.* Nouvelle Librairie Nationale, Paris (second edition, 1924, 425p.) (1922).
TERZAGHI, K. - Stress conditions for the failure of saturated concrete and rock. *Amer. soc. Testing Material Proc.* 45, 777 - 792 (1945).

TERZAGHI, K. - Mechanics of landslides. In *Application of Geology to Engineering Practice* (Berkeley vol) Paige, Sidney, ed. *Geol. Soc. Amer.*, 83 - 123 (1950).
THOMSON, R.I. - The nature and significance of large 'blind' thrusts within the northern Rocky Mountains of Canada. In *Thrust and nappe tectonics*, Special volume of the Geological Society of London **9**, 449 - 462 (1981).
TIKOFF, B. and FOSSEN, H. - Simultaneous pure and simple shear: the unifying deformation matrix. *Tectonophysics* **217**, 267 - 283 (1993).
TONDJI BIYO, J-J. - *Chevauchements et bassins compressifs, influence de l'érosion et de la sédimentation: modélisation analogique et exemples naturels.* Thèse Université Rennes, 412p. (1993).
TÖRNEBOHM, A.E. - Om fjallproblemet. *Geol. Foren. Stockholm forh.* **10**, 328 (1888).
TÖRNEBOHM, A.E. - Grunddragen af det Centrala Skandinavians *Bergbyggnad kongl. Svenska Vet. akad. Handl.*, Bd. **28**, number 5, 190 - 195 (1896).
TOWNSEND, C. - Thrust transport directions and thrust sheet restoration in the Caledonides of Finnmark, North Norway *Journ. Struct. Geol.***9**, 3, 345 - 352 (1987).
TREVISAN, L. - Nuovi orientamenti dello studio della tettonica. *Historia Naturalis* **1**, number 3, Roma (1946).
TRUMPY, R. - The Glarus nappe: a controversy of a century ago. In *Controversies in modern geology*, edited by D. W. Muller, J. A. McKenzie and H. Weissert, Academic Press limited, 385 - 404 (1991).
UMBGROVE, J.H.F. - The roots of the Alps. *Konink Nederlands Akad Van Wetenschappen*, CI, number 7 (1948).
VAN BEMELEN, R.W. - The undation theory of the development of the Earth's crust. *Proc. 16th Intern. Geol. Congr. Washington, D.C.*, V.L., 965 - 982 (1933).
VIALON, P., RHULAND, M. and GROLIER, J. - *Elements de tectonique analytique.* Masson, Paris (1976).
VIALON, P. BONNET, J.L., GAMOND, J.F. and MUGNIER, J.L. - Modélisation des déformation d'une série stratifiée par le déplacement horizontal d'un poinçon. Application au Jura. *Bull. Soc. Geol France* XXVI, number 1, 139 - 150 (1984).
VOIGHT, B. - *Mechanics of thrust faults and decollement.* Edited by B. Voight, Benchmark Papers I Geolgoy, vol. 32, Dowdon, Hutchinson and Ross, Inc., Stroudburg, Pennsylvania (1976).
WEGENER, A. - Die entstehung der kontinente. *Geologische Rundschav*, **3**, 276 - 292 (1912).
WILKERSON, M.S. - Differential transport and continuity of thrust sheets. *Journ. Struct. Geol.***14**, 6, 749 - 751 (1992).
WILSCHKO, D.V. - A mechanical model for thrust sheet deformation at a ramp *J. Geophys. Res.* **84**, 1091 - 1104 (1979).
WILTSCHKO, D.U. - Thrust deformation at a ramp: summary and extensions of an earlier model. In *Thrust and nappe tectonics*, Special Volume of the Geological Society of London **9**, 55 - 63 (1981).
WILLIAMS, G.D. - Rotation of contemporary folds into the X direction during overthrust processes in Laksefjord, Finmark. *Tectonophysics* **48**, 29 - 40 (1978).
WILLIAMS, G.D. and CHAPMAN, T. - Strain developed in the hangingwalls of thrusts due to their slip/propagation rate: a dislocation model. *Journ. Struct. Geol* **5**, 6, 563 - 571 (1983).
WILLIAMS, G.D., CHAPMAN, T. and MILTON, N.J. - Generation and modification of finite strain patterns by progressive thrust faulting in the Laksefjord nappe. *Tectonophysics* **107**, 177 - 186 (1984).
WILLIS, B. - Mechanics of Appalachian structure. *U.S. Geol. Surv. 13th Ann. Rept.*, Part II (1893).
WOOD, D.S. - Patterns and magnitudes of natural strain in rocks. *Phil. Trans. R. Soc. London* **27**, 373 - 382 (1973).

WOODWARD, N.B., GRAY, D.R. and SPEAR, D.B. - Including strain data in balanced cross section. *Journ. Struct. Geol.* **8**, 313 - 324 (1986).

ZOETEMEIJER, R. and SASSI, W. - 2D reconstrution of thrust evolution using the fault bed method. In *Thrust tectonics* (ed. McClay), Chapman and Hall, 133 - 140 (1992).

Index

Aar 124
Acceleration 48
Active folding 105
Aiguilles Rouges 126
Algeria 56
Allochthon (definition) 1
Alps 37,38,43,105,126,127
        Provence 36
        southern 132
        western 105,110
        Swiss 3,38,59
        Austrian 33
Analogue experiments 71,90
Analogue materials 40,43
Angle of internal friction 48
Anhydrite 55
Anticline 11
Appalachians 29,67,106
Autochthon (definition) 1
Austria 93
Axis
        shortening 98,119
        stretching 101,119,120
Backthrust 14,18
Balancing 25,26
Basal layer 7,18
Basin
        foreland 128,129

flexural 128
French-Belgian coal 30,35
molassic 124,128,130
Batholith 59,70
Bearpaw Mountains 59,71
Blind thrust 14
Borderline cases 52,79
Boudinages 5
Bow and arrow rule 116,117,130
Brittle 45,56,64,67,73,76,77,115
Caldera 70
Canada 36
Carletonville 58
Carpathes 31
Coaxial 82,89,109,110
Coefficient of internal friction 48,67,74
Coesite 125
Cohesion 48,52,53,56,67,74
collapse 60,62
Competent 91
Compression 33, 42, 43, 46
rear 65,77,85,87,89,92,93,99,106,107,119,120,133
Conditions
of stability 56
boundary 82,9,109
Conservation
of surface 24
of volume 24
Constriction 98,99,101,121
Continental Drift 39,44
Convection 44
Cover substitution 18
Criteria
Navier-Coulomb 47,50,52,53
Critical wedge 64, 65, 66, 67, 68, 69, 77
Crustal thickening 128,129
Curvimetry 25
Cylindricality 110
Decollement level 11, 54, 57, 58, 64, 67, 124, 130, 132
Deformation
continuous 54
discontinuous 56
incremental 104,105
pseudo-viscous 59,67,77
Dehydration 55
Depocentres 128

Index 153

Detachment fold 107
Diagenesis
Diapir 69,81
Digne 3,16
Dislocation 53
Displacement 30,33,35,36,41,43,44,50,53-61,64-77,79,82,87-91,98-107,113-119,128
    differential 118,130
    radial 87,99
Distortion 82
Diverticulation 45
Ductile 35,43,46,53,56,73,82
Ductility 5,7,18,20,22
duplex 15
Eclogite 125
Eigenvalues 86
Eigenvectors 86
Elastic 56,70
Ellipsoid 98-101
    strain 86,91,98-101,119
Emplacement mechanism 40-47,77,82,89,106,127,133
Encapuchonnement 18
Erosion 32,73-76,115,128
Evaporites 126
Extruding-Spreading 83,84,87-88,98-201,126
Factorisation (of strain) 84,88,91
Fault
    antithetic 14
    conjugate 114,115
    reverse 29,30,54
    strike-slip 118,128
Fault-bend fold 14
Fault-propagation fold 14
Fibres 104-105
Finland 21
Flat 76 108,110-113,133
Flattening 90,98,101,121
Flexural
    basin 128
    -flow fold 112
Flexure (lithospheric) 111,128
Flinn (parameter) 91,98
Floor thrust 14
Flow law 57,58,63
Flowing 35,44,58,60,64,77,79,83,91,109
    viscous 44
Fluid push model 69

Flysch a Helminthoides 105,110
Fold and thrust belt 66,90,129,130
Folds
    active 105,109
    box 67,90
    detachment 107
    fault-bend 107
    fault propagation 107
    -nappes 1,18,22
    passive 108,110
    sheath 106,110
    upright 90,106
foliation 33,124
Forces
    surface 41
    volume 41
    tectonic 41
foreland
    basin 128,129
    dipping duplex 15
Friction 48,51,67,74,84,130
Frontal rolling 91,94
Gavarnie 102
Geosyncline 44
Glaciers 45,60,63,87,93,109
Glarus 35,36,40,59
Gliding
    ductile 79,82,84-85,91,118,121,126
    gravitational 59,60,70,72,74,75,77,79,103,105,130
    rigid 79,82,126,128
Gouges 7
Gradient
    strain 82,89,91,96,97
    displacement 92
Granite 55,70
Graz 93
Greenschists 125
Gypsum 51,55,56
Helvetic 99-101,123,125-130
Hinterland dipping duplex 15
Horse 15
Ice 109
Idaho 59
Imbricate structure 14
Imbrications 64,129
Incompetent 10

Instability
    gravitational 60
    mechanical 105
Interface 109
Intrusion 59,69-71
Isopaques 124-130
Jura 31,67,90,11,123-126,129-132
Kinematics (approach) 45,70,76,78,81,82,109,123,132,133
Kink band 116
Klippe 36,124
Keystone Muddy Mountain 70,76
Laksefjord 21
Lenses 5
Listric 8
magma 59,69-71
Marble 58
Marnes 8
Mechanical effects 110
Microstructures 34
Models
    analogue 40,90
    emplacement 85,100
    reduced scale 43,63,77,113
Mohr-Coulomb 48,54,67,68,70,73,77
Molasses 128,129
    sub-Alpine 129,130
Molassic plateau 129,130
Montana 59
Montagne Noire 102
Mt Blanc 124,126
Mt Perdu 102
Morcles 100,101,126
Movements
    divergent 96
    radial 96
Nappe (definition)
    basement 123,125
Nevada 70,76
Newtonian 56,58,59
Olistoliths 129
Olistostromes 43,129
Ophiolite 6
Palaeogeography 102,103
Palaeomagnetic 45,103

Paradigm 45
Parautochthon 17
Piggyback 13
Pin-line 28
Plate
    Adriatic 123,124
    African 123
    European 123,124,125
    tectonics 39,45
Plastic 35,51,56,58,60,64,66-68,72,77
Po (Plain) 133
Pre-Alps 37,38,43,123,124,126-128
Pressure
    pore fluid 44,51,52,74
    hydrostatic 52,69,70
    isotropic 52
    solution 63
Pyrenees 102
Quadratic elongation 86
Ramp 76,108,110-120
    emergent 70,73-76,108,110
    frontal 75,115
    lateral 116,119,120
Rates
    strain 58,63,86
    displacement 57,58,73,76
    erosion 73
    subsidence 128
Regime
    strain 82,95,110
    coaxial 89,110
    non-coaxial 110
Reverse metamorphism 7
Rheology 45,51,56,60,63,67-69,72,76-77,109
Rigid translation 84,91,102,103,105,115
Rockies 36,60,61
Roof thrust 15
Rotation 103-105,112,116,119-120
Ruby's Inn 70
Sand 43,56,67,0,76,85,115
Scandinavia 36
Schistosity 45,46,82,86,87,90-92,94-95,98-104,111-112,119-121
Scotland 30,34,38

Shear
    pure 85,8,8891,95-9,103,113,118,121
    simple 57,85,87,89-92,95-98,103,106-107,111-113,118
Sierra Nevada 70
Silicone 70-71,76,96
Snow 65,109
Sole thrust 14
South Africa 58
Spreading 77,79,83,84,87,92-94,98-102,104,119,121
    gravitational 44,60,63-64,66,68,77, 85,87,95,96,100,102,106
    radial 87,96
Stereogram 119
Strain
    factorisation 84,91
    partitioning 90,132
    pattern 78,81,85,100,106
Stream lines 109
Stretching
    concentric 88,96-100
    incremental 104,105
    lineation 33,91,101-104,106,118
    radial 97-100
Striae 103
Subduction 69
Tectonic slice 17
Tectonic outlier 17
Tensor
    Finger 85,86
    strain rate 86
Theories
    elasticity 54
    deformation 46
    fixist 42
    mobilist 42,44,46
    orogenic 42
    verticalist 42,44,45
Thick-skinned thrusting 1,5
Thin-skinned thrusting 1,2,5
Threshold
    plasticity 66
Thrust (definition)
Thrusts
    blind 14
    conjugate 67,113-115
    crustal 103,118,123-126,128,133

        imbricate 14
        outlier 17
        surface 1
        thin-skinned 1,5
        thick-skinned 1,2,5
Tip line 14
Topographic dome 59,62,70
total area 26
Trajectories
        displacement 102,103,104
        schistosity 82,86,90,92,95,96,99,100,101
        stress 54
        particle 109,119,120
Transgression 33,42
Transport direction 103,119
Triangle zone 14

Unfolding 25
Utah 70

Vectors
        displacement 59,96,97,102,104
Veins 104-105
Vertical velocity profile 59,109
Virginia 106
Viscosity 56109,110
Visco-plastic 60
Volcanic arc 69,70
Wedge shape 10
Window 17
Wrenching 118
Zone
        pressure shadow 104-105,113
        decollement 56-58,90,106
        external 69,81,102,124-129
        Helvetic 123,128
        internal 43,69,123-128,132
        triangle 14

LEGEND FOR GEOLOGICAL CROSS-SECTIONS

KEY	NAME	REFERENCE
	Sedimentary Rocks	Fig. 1 Fig. 2 Fig. 3 Fig. 4 Fig. 5 Fig. 7 Fig. 8 Fig. 9 Fig. 10 Fig. 11 Fig. 12 Fig. 29 Fig. 39 Fig. 41 Fig. 65 Fig. 66 Fig. 68 Fig. 69
	Undeformed Granite	Fig. 9 Fig. 41 Fig. 66 Fig. 68 Fig. 69
	Orthogneiss or Paragneiss	Fig. 12 Fig. 41 Fig. 65 Fig. 66 Fig. 69
	Continental Crust	
	Lithospheric Mantle	Fig. 6